BRIAN DAVID KERL, PH.D.

COMMAND CLIMATE AND ETHICAL BEHAVIOR

PERSPECTIVES FROM THE COMMANDANTS OF THE MARINE CORPS

POTOMAC INSTITUTE FOR POLICY STUDIES

Published, 2019 by the Potomac Institute Press

ISBN: 978-1-7337685-2-8

Cover image photo by Alex Taliesen.

Potomac Institute Press
Potomac Institute for Policy Studies
901 N. Stuart St, Suite 1200
Arlington, VA, 22203
www.potomacinstitute.org
Telephone: 703.525.0770; Fax: 703.525.0299
Email: webmaster@potomacinstitute.org

CONTENTS

CHAPTER FOUR

CHAPTER FIVE

APPENDIX A

APPENDIX B

APPENDIX C

APPENDIX D

APPENDIX E

Acknowledgments

There are many people I wish to thank for making this project possible. First, I would like to thank the eight Commandants of the United States Marine Corps who graciously gave their time to be interviewed and to review their respective sections of the dissertation. I would also like to thank the people at the Marine Corps University and the Lejeune Leadership Institute who provided me access to certain interviews on the base at Quantico, and who offered their support throughout the research process. I would like to thank Alex Taliesen from the Potomac Institute for Policy Studies and Sherry Loveless for their assistance and patience.

I would like to thank the four members of my dissertation committee. Dr. Fred Galloway was my academic advisor during all five years of my doctoral education at the University of San Diego and served as committee chair. Fred was always available and he continually provided great advice on how to approach the project. I'll never forget our conversations while listening to classic rock in his office. Dr. Lea Hubbard taught the Advanced Qualitative Analysis course that prepared me for this research project. Lea offered sage advice on how to tackle the transcripts, analytical memos, and the jot notes necessary for the data collection phase of the study. Dr. Bob Donmoyer taught me how to write. It was in his basic Qualitative Analysis course that I really started to appreciate and learn the art and science of good writing. Bob also taught me the fine points of how to conduct personal interviews, which came in handy for the interviews with the Commandants. I would also like to thank Dr. Paolo Tripodi for providing a Marine Corps perspective to the dissertation committee and for believing in the project. He was a great friend throughout the process.

Finally, I want to thank my family. I could not have finished the program or conducted the research without the support of my wonderful wife,

Sally Cohenour-Kerl, and our son, Brian. They both made sacrifices while I spent countless hours at Copley Library, or in one of the classrooms at the School of Leadership and Education Sciences. I would also like to thank my parents, Dave and Lucy Kerl, for their love and support, as well as my siblings, Lisa Ray and Brad Kerl, who had to put up with my stories about school and my research project.

Foreword

I was still a young man when my father, Victor H. "Brute" Krulak—a Brigadier General who eventually became a Lieutenant General, USMC—was asked by then-Commandant of the Marine Corps, General Randolph McC. Pate: "Why does the U.S. need a Marine Corps?"

In response to the question posed by General Pate, my father's answer was simply that the U.S. "does not need a Marine Corps." However, he went on to explain that due to its history of winning battles, its unfailing preparedness, and its immediate availability to confront any foe, in any clime and place; America wants a Marine Corps. He then went on to explain the most important reason why America wants a Marine Corps. He simply stated, "We exist today—we flourish today—not because of what we know we are, or what we know we can do, but because of what the grassroots of our country believes we are and believes we can do. The American people believe that Marines are downright good for the Country; that the Marines are masters of a form of unfailing alchemy which converts unoriented youth into proud, self-reliant, stable citizens—citizens into whose hands the nation's affairs may safely be entrusted—and likewise, should the people ever lose that conviction—as a result of our failure to meet their high, almost spiritual, standards, the Marine Corps will quickly disappear."

Americans expect their Marines to be more than the finest fighting force in the world. They expect their Marines to act with integrity; to demonstrate the Corps' values of honor, courage, and commitment on a daily basis; and to epitomize everything that is good about our nation by personifying the ideals upon which it was founded. This high expectation Americans have for their Marines is, truly, almost "spiritual" in nature.

Yet, throughout its illustrious past, the Marine Corps has had a small percentage of its force that has engaged in unethical behavior. Their behavior has tarnished the Corps' image through deeds of misconduct. Such misconduct has caused the American public and the Congress to, at times, be amnesiac about the sacrifices Marines have made for over 243 years. It has caused some to ask the question, "Are the Marines teaching my son or daughter values?"

As the 31st Commandant of the Marine Corps, I can unequivocally state in response to the question above, yes, the Marine Corps teaches values to all Marines and that a Marine's character remains the bedrock upon which everything else is built! This idea is central to what we call the Transformation—the ongoing process that begins with the recruiter and continues throughout the Marine's life and ends with that Marine returning to society as a better citizen for having served in our Corps. The Transformation has many phases, but the most critical phase and most relevant to this book is the sustainment phase—the responsibility of the unit leader.

It is the commanding officer's responsibility to establish a strong command climate focused on the character development of the individual Marine through strengthened moral fitness that emphasizes the Marine Corps' core values. Critical to the Transformation is the sustainment process where the values that have been embedded in new Marines at boot camp and Officer Candidates School are reinforced on a routine basis.

In his book, Brian Kerl, a retired colonel of Marines who has served on active-duty for 28 years in both command and staff positions, has captured the most important thing a commanding officer is charged with—establishing a command climate which facilitates the on-going development and sustainment of the individual Marine's character. The book covers this idea from the perspectives of eight Commandants of the Marine Corps. Each Commandant discusses the concept and the critical role played by the command climate and how it influences the ethical behavior of the people in the organization. When Brian interviewed me,

I told him that his topic was critical to the survival of the Marine Corps. The Corps is never questioned by Congress or the American people for its warfighting prowess. However, during my tenure as Commandant and for many of my peers, incidents of misconduct caused from situational ethics, or situational morality have required our Corp's senior leadership to explain why these incidents occur and what the Corps is going to do to fix the problem.

One solution to maintaining the trust of the American people and the Congress is for commanders to focus on the sustainment phase of the Transformation. As General Lejeune said, "...especially commanders, are responsible for the physical, mental, moral welfare, as well as the discipline and military training of the men under their command who are serving the nation in the Marine Corps..." Yes, the Corps' warfighting ability is important to the nation, but what will keep the Corps in existence for the long haul is the notion that "Marines are downright good for the country." The American public continues to believe that Marines are individuals of superior character and virtue to "whom the nation can entrust its affairs." Commanders must establish a command climate that is focused on the character and the ethical behavior of the Marines in the command. This book will shed light on how eight Commandants have approached this critical task.

Charles C. Krulak
General, U.S. Marine Corps (Ret.)
31st Commandant of the U.S. Marine Corps
Former Member of the U.S. Joint Chiefs of Staff

Photo by mj0007.

Preface

Throughout my service in the Marine Corps, many role models—both enlisted and officer—emphasized to me the importance of integrity and character. These Marines suggested that integrity is the most crucial attribute for a Marine to possess. As a commanding officer, I made every attempt to focus on the character development of my Marines and to emphasize the Marine Corps' values of honor, courage, and commitment.

During my 28-year service, I witnessed good command climates and endured poor command climates. The good climates promoted a positive environment where people were happy and there was solid unit cohesion. The poor climates made people want to get out of the Marine Corps. The units with poor command climates experienced numerous disciplinary problems, which degraded unit readiness and hindered mission accomplishment.

The United States Armed Forces has recently experienced numerous policy changes that reflect developments in contemporary society, including women serving in combat arms occupations (e.g., the infantry), transgender personnel serving in the military, and the extensive use of social media. These changes have underscored the significance of good command climates. A good command climate focuses on the Marine Corps' core values of honor, courage, commitment, and character development. Such climates produce better people, who typically make better Marines. But, unfortunately, on numerous occasions, a minority of Marines have tarnished the image of the Marine Corps through their misconduct. The past few years have revealed several articles and headlines in the news regarding the poor conduct of United States Marines. It was difficult in 2017 to watch the Commandant of the Marine Corps testify before Congress in response to the social media scandal known as "Marines United."

These incidents and my personal convictions and experiences as a former commanding officer led me to initiate the research project on command climate that is described in this book. In light of the instruction provided to all Marines and the emphasis on the Marine Corps' values, I was curious as to why some Marines continue to take the wrong path. To better understand this crucial subject, I leveraged the wisdom, experience, and passion of eight Commandants of the Marine Corps. I sought and analyzed the perspectives of these leaders on a number of questions—For example: How does the command climate determined by the commanding officer influence the ethical behavior of Marines? What does the right command climate look like? How does a commander establish the right command climate?

The American people's expectations of the Marine Corps are extremely high. Yet, for some reason, some Marines cannot comply with the Marine Corps' high standards and live up to the expectations of the American people. It is the responsibility of all Marines, especially the commanders, to make every effort to meet the high expectations Americans have for their Marines. Setting the right command climate that promotes ethical behavior will enable continued success for the U.S. Marine Corps in its service to the American public. It is my hope that this book enlightens current and future leaders on the definition, establishment, and development of command climate and the role it plays in influencing ethical behavior within U.S. Marine Corps, other military services, and civilian organizations.

Introduction

The history of the human race is about the increasing use of data to generate increasing levels of intelligence. Whether through evolutionary changes, oral traditions, written texts, or computer records, humankind has used experiences resulting in sensory inputs to accumulate increasing abilities to obtain higher levels of intelligence and to thereby adapt and manipulate the human environment.

For 243 years, the United States Marine Corps has displayed the ability to adapt to changes in the strategic environment, build on its rich heritage, and win on the modern battlefield.[1] Throughout its history, the Corps has developed leaders who possess the ability to make Marines, to train them to fight and win our nation's battles, and eventually to return quality citizens back to society. Not surprisingly, the Marine Corps invests heavily in leadership instruction for its commanders. As part of this instruction, the Marine Corps educates and socializes its commanding officers on ethical behavior, which is at the center of its ethos.[2] For example, the Marine Corps War College (MCWAR) offers multiple courses of ethics instruction and includes ethics materials in several other aspects of the curriculum.[3] In addition, the Director of the ethical instruction program conducts strategic-level discussions (often with national implications) that regularly include general officers as guest participants.

Current conditions provide significant tests for the Marine Corps' leadership development process. Today's commanding officers face complex challenges: the wars in Iraq and Afghanistan, strategic uncertainty,

1. Department of the Navy, Headquarters Marine Corps. *Marine Corps Strategy* 21. Washington, D.C: U.S. Government Printing Office, 2000.
2. Department of the Navy, Headquarters Marine Corps. *Marine Corps Warfighting Publication* 6-11, *Leading Marines*. Washington, DC: U.S. Government Printing Office, 2014.
3. Behn, B. U.S. Air War College. "The Stakes are High: Ethics Education at US War Colleges." Graduation Thesis, Air University, 1-27. Maxwell AFB, AL: Air University Printing, 2016.

societal changes that carry over to the military such as transgender personnel, and the extensive use of social media. They must also—as they always have—contend with the perennial facts of human frailty, which are inevitably exacerbated by the stressors unique to the military, such as the lengthy separation from loved ones and multiple relocations of duty station.[4] Some of these challenges have relatively-easy-to-figure-out technical solutions, while others have no standardized solution. Heifetz calls these latter, more complex problems, "adaptive challenges."[5] The importance of command climate, particularly its ethical dimension, is demonstrated by the diverse range of adaptive challenges that Marine commanders must deal with both at home and abroad.

Ethical climate theory[6] analyzes the types of organizational climates that guide employees' ethical decision making and behavior.[7] Ethical climate theorists define ethical climate[8] as "shared perceptions regarding organizational policies, practices, and procedures that emphasize ethical content."[9] Key features of an organization with an ethical climate include humility, no tolerance for destructive behaviors (e.g., incivility, aggression, discrimination, or sexual harassment), justice, integrity, trust, a focus on process (how organizations achieve their goals), structural reinforcement (rewards, employee evaluations, and worker decision making rights), and social responsibility.[10]

---

4. Department of the Navy (DON), Headquarters Marine Corps (HQMC). *Marine Corps Strategy* 21.
5. Heifetz, R., and Linsky, M. *Leadership on the Line: Staying Alive through the Dangers of Leading*. Boston, Massachusetts: *Harvard Business Review* Press, 2002.
6. Simha, A., and Cullen, J. "Ethical Climates and Organizational Outcomes: Implications from the Past and Prophecies for the Future." *Academy of Management Perspectives* (Nov 2012): 20-34; Victor, B., and Cullen, J. B. (1987). A theory and measure of ethical climate in organizations. *Research in Corporate Social Performance and Policy* 9, 51-71; and Victor, B., and Cullen, J. B. (1988). The organizational bases of ethical work climates. *Administrative Science Quarterly* 33, 101-125.
7. Martin, K., and Cullen, J. "Continuities and Extensions of Ethical Climate Theory: A Meta-Analytic Review." *Journal of Business Ethics* 69 no. 2 (2006): 175-194.; Schminke, M., Arnaud, A., and Kuenzi, M. "The Power of Ethical Work Climates." *Organizational Dynamics* 36 no. 2 (2007): 171-186.
8. See Chapter 1 for climate theory, definitions, and terms.
9. Victor and Cullen, 1987, 1988, p. 101-125.
10. Johnson, C. *Meeting the Ethical Challenges of Leadership, Casting Light or Shadow*. Thousand Oaks, CA: SAGE Publications, Inc, 2015.

An organization with an ethical climate will foster greater creativity, benevolence, and commitment.[11] Leaders who establish ethical climates will promote organizations with higher morale, lower employee turnover, and positive outcomes based on trust and shared ethical principles.[12] Also, leaders working within those climates will be more predisposed to ethical decision making and have a greater awareness of ethical blind spots (e.g., careerism, corruption, etc.) that can negatively impact organizational performance.[13]

The Marine Corps commonly refers to the organizational climate as "command climate" and that is the term used throughout this book. Both ethics and command climate are important to the Marine Corps Values Program, which directs commanders to "integrate Marine Corps values training into organizational training plans."[14] The Marine Corps' core values of honor, courage, and commitment provide the ethical framework for all Marines.[15] An ethical climate supports the continuity of core values training and reinforces trust between the leaders and subordinates critical to leading Marines.[16] Additionally, an ethical climate might reduce or mitigate incidents and problems such as domestic violence, driving under the influence, suicides, and the inappropriate use of social media.[17] By establishing a command climate focused on encouraging ethical behavior, the commander empowers Marines to act ethically on their own, thus reducing the number of incidents associated with misconduct and enabling the commander and his staff to focus on training and combat readiness.[18]

11. Simha, A., and Cullen, J. "Ethical Climates and Organizational Outcomes: Implications from the Past and Prophecies for the Future." *Academy of Management Perspectives* (Nov 2012): 20-34.
12. *Ibid.*
13. Bazerman, M., and Tenbrunsel, A. *Blind Spots: Why We Fail to Do what's Right and What to Do about it.* Princeton, NJ: Princeton University Press, 2011.
14. DON, HQMC, 1996, *Marine Corps Values Program*, p. 4.
15. DON, HQMC, 2014, *MCWP 6-11, Leading Marines.*
16. Katolin, D. "Ethical Development: How We Shape Our People's Behavior." *Marine Corps Gazette* 100 no. 9 (2016a): 53-57.; McAleer, P. "Our Corps' Soul at Stake." *Marine Corps Gazette*, 101 no. 6 (2017): 47-48.
17. Katolin, D. "Ethical Development: How We Shape Our People's Behavior."; Doty, J., and Gelineau, J. "Command Climate." *Army Magazine*, July 2008, 22-24; Weber, J., and Gerde, V. "Organizational Role and Environmental Uncertainty as Influences on Ethical Climate in Military Units." *Journal of Business Ethics* 100 (2010): 595-612.
18. Katolin, D. "Ethical Development: How We Shape Our People's Behavior."; Olsthoorn, P. *Military Ethics and Virtues.* New York, NY: Routledge, 2011.

Retired General Charles Krulak[19] points out that "the leader [...] must clearly demonstrate the true underpinning of his moral authority—his unquestionable character."[20] A critical component of a climate focused on ethical behavior is the leadership quality of the commanding officer.[21] The commanding officer, as the senior person leading the organization, is responsible for establishing the right command climate based on non-negotiable values that guide the organization in everything it does.[22] An ethical climate relies heavily on the ethical guidance of the commander,[23] whose influence is based on integrity, values, and ethics.[24]

At the same time, the command climate depends on the network of personalities within the organization.[25] In the Marine Corps, commanding officers change positions every two years. The command climate can be described as the "feel" of the organization. It involves internal aspects of the organization such as attitudes, loyalties, motivations, and perceptions from its members, who change duty assignments on a routine basis.[26]

Despite all the conferences, studies, and the positive changes implemented by the Marine Corps, ethical problems continue to plague Marine commanders. No more evidence is needed than the social media scandal "Marines United," which involved the posting on social media of nude photographs of female Marines accompanied by lewd captions.[27]

---

19. Krulak, C. *Professional Reading Program; Commandant's Choice for 1995*. ALMAR 157/95, 1995.
20. Freedman, D. *Corps Business: The 30 Management Principles of the U.S. Marines*. New York, NY: Harper Collins Publishers, 2000, p. xii.
21. Katolin, D. "Military Ethics: How Maneuver Warfare Demands Adherence to Our Values." *Marine Corps Gazette* 100 no. 8 (2016b): 44-46.; Johnson, R. "How Can Leaders Maintain Ethical Command Climates?" *LTX Magazine* 3 no. 2 (2013): 1-3.
22. Doty, J., and Gelineau, J. "Command Climate." *Army Magazine*, July 2008.
23. Miller, C., and Poole, E. *Ethical Leadership, Global Challenges and Perspectives*. New York, NY: Palgrave MacMillan, 2011; Kirkpatrick, S. A., and Locke. E.A. "Leadership: Do traits matter?" *Academy of Management Executive* 5 no. 2 (1991): 48–60; Katolin, D. "Military Ethics: How Maneuver Warfare Demands Adherence to Our Values." *Marine Corps Gazette* 100 no. 8 (2016b): 44-46.
24. Pfeffer, J., Salancik, G., and Leblebici, H. *The Effect of Uncertainty on the Use of Social Influence in Organizational Decision Making*. Lexington, KY: The University Press of Kentucky, 1976; Heifetz, R. *Leadership Without Easy Answers*. Cambridge, Massachusetts: Harvard University Press, 1994.
25. Department of Defense (DoD), FM 22-100, *Army Leadership*, 1999.
26. Bullis, C. and Reed, G. U.S Army War College. *Assessing Leaders to Establish and Maintain Positive Command Climate*. Department of Command, Leadership and Management Study, i-71. Carlisle, PA: Carlisle Barracks Printing, 2003.
27. McAleer, P. "Our Corps' Soul at Stake." *Marine Corps Gazette*, 101 no. 6 (2017): 47-48.

Another recent incident occurred at the Marine Corps Recruit Depot, Parris Island, where a Marine recruit—who ultimately committed suicide—was hazed by being placed inside a clothes dryer because of his Muslim faith.[28] Finally, the Glenn Defense Marine Asia scandal, known as "Fat Leonard," implicated both Navy and Marine Corps leaders in unethical conduct dealing with civilian commercial contracts.[29] These examples from the news headlines, as well as other incidents of ethical misconduct, have the potential to destroy unit cohesion, ruin morale, and ultimately degrade the trust and confidence of the American people in their Marines.[30]

The purpose of my research was to learn about the role command climate plays in a Marine Corps commanding officer's ability to influence ethical behavior. The study provided an opportunity to examine this question through the perspectives of eight Commandants of the Marine Corps (CMC), including the current Commandant, and other key Corps personnel such as former commanding officers and ethics instructors. Prior to interviewing the CMC, I examined vital components of the Marine Corps' officer development programs, including the Senior Service College and the Cornerstone Course taught at the Lejeune Leadership Institute (LLI) for Marine Corps commanding officers. I requested the assistance of former commanders and ethics instructors in the development of the questions that were used to interview the CMC (see Appendix C). This preliminary work helped inform the interviews with the CMC and furthered my understanding of the influence of command climate relative to ethical behavior within a Marine Corps organization. Please see Appendix A for a complete description of the research methods, data analysis and coding strategy used for the study in addition to information on the 2016-17 pilot study that informed the research protocol.

28. Harkins, G., and Schogol, J. "Marines find culture of hazing, abuse at boot camp after recruit's death at Parris Island." *Marine Corps Times*, September 8, 2016.
29. San Diego Union Tribune. "9 More Navy Officers Indicted." *San Diego Union Tribune*, March 15 2017b.
30. McAleer, P. "Our Corps' Soul at Stake." *Marine Corps Gazette*, 101 no. 6 (2017): 47-48.

According to the CMC, the Corps does a thorough job of training and educating its Marines in the importance of doing what is right, but there is always room for improvement. All Marines receive detailed instruction on core values, standards, the ethos of the Marine Corps, and orders and regulations during entry-level training, as well as during follow-on professional schooling (e.g., the Sergeants School, the Expeditionary Warfare School for captains).

Most Marines serving the nation do so honorably and abide by the Marine Corps' core values and meet its high standards. Based on the history of the United States Marine Corps and its reputation for discipline, esprit de corps, and professionalism, it is reasonable to assume that most commanders establish the right climate for their Marines. However, incidents of misconduct involving a small number of Marines continue to plague commanders. As mentioned previously, some of these incidents make the headlines of the local and national news, which may slowly erode the trust and confidence of the American people in their Marine Corps. If the Marine Corps always strives to improve and is willing to be self-critical, making the necessary adjustments to continue to be the most lethal and capable warfighting organization on the planet, then it should welcome the results of this study and examine how its commanders establish their command climates.

I hope that this book contributes to the Marine Corps' leadership development programs by providing recommendations, based on the CMC's perspectives, to assist with the training and education of commanding officers and other leaders. It builds on the work already initiated by the LLI and other Marine Corps Training and Education Command initiatives. The goal is to encourage ethical climates designed to enhance combat effectiveness and promote character development within their organizations, and to offer data that will help inform subsequent Marine Corps studies on command climate and leadership.

CHAPTER ONE

Background—Environment, Theoretical Framework, and Ethical Leadership

A commanding officer is "a military officer in charge of a unit, post, camp, base, or station."[31] As commanding officers, men and women in command of Marines have the "unmitigated personal responsibility and accountability" toward the people under their care and might give orders that send their Marines into combat.[32] The difficult decisions facing commanders in today's environment require high degrees of maturity, discernment, and prudence.[33] Often, young Marines are tasked with making difficult moral decisions, which magnifies the need for a command climate focused on ethical behavior.[34] The recent wave of military scandals and unethical conduct by military personnel has brought renewed attention to the role of ethics in military organizations.[35] The absence of a command climate focused on ethics may be a contributing factor to breakdowns in behavior and discipline.[36]

31. *American Heritage Dictionary*, 2013.
32. Couch, K. (2016). "Officership and Leadership." *Marine Corps Gazette* 100 no. 6 (2016): 22-26.
33. Freedman, D. *Corps Business: The 30 Management Principles of the U.S. Marines*. New York, NY: Harper Collins Publishers, 2000.
34. Freedman, D. *Corps Business: The 30 Management Principles of the U.S. Marines*.
35. *San Diego Union Tribune*. "Marine Corps Investigating Nude Photo Scandal." San Diego Union Tribune, March 7 2017a; *San Diego Union Tribune*. "9 More Navy Officers Indicted." *San Diego Union Tribune*, March 15 2017b; McAleer, P. "Our Corps' Soul at Stake." *Marine Corps Gazette*, 101 no. 6 (2017): 47-48.
36. Martin, K., and Cullen, J. "Continuities and Extensions of Ethical Climate Theory: A Meta-Analytic Review." *Journal of Business Ethics* 69 no. 2 (2006): 175-194; Myer, A., Thoroughgood, C., and Mohammed, S. "Complementary or Competing Climates? Examining the Interactive Effect of Service and Ethical Climates on Company-Level Financial Performance." *Journal of Applied Psychology* 15 no. 3 (2016): 1178-90.; Katolin, D. "Ethical Development: How We Shape Our People's Behavior." *Marine Corps Gazette* 100 no. 9 (2016a): 53-57; Rowell IV, G. *Marine Corps Values-Based Ethics Training: A Recipe to Reduce Misconduct*, Strategy Research Project, 1-29). United States Army War College. Carlisle, PA: Carlisle Barracks Printing, 2013; Mayer, D. "A Review of the Literature on Organizational Climate and Culture." *The Oxford Handbook of Organizational Climate and Culture* (2014): 1-28.

This chapter provides a brief discussion on the complex environment in which commanding officers operate, and how the Marine Corps is educating senior military commanders to lead ethically within this environment. It also covers an overview of the theoretical framework on ethical climate, definitions of ethical climate most relevant to the study, and ethical leadership. Finally, some background on the role of the commander and examples of ethical blind spots that may hinder a commander's ability to maintain a command climate focused on ethical behavior is discussed.

The Environment

The importance of an ethical command climate cannot be understated given the range of challenges that Marine commanders must contend with at home and abroad.[37] Keenan suggests this environment includes domestic problems, such as suicides, sexual assaults, and substance abuse, as well as the potential issues associated with the absence of an ethical climate.[38] Whiffen argues that the Army needs leaders who have "the ability to modify individual and collective actions based on circumstances."[39] This argument correlates with Heifetz and Linsky's description of adaptive challenges whereby "without learning new ways—changing attitudes, values, and behaviors—people cannot make the adaptive leap necessary to thrive in the new environment."[40]

These perspectives are relevant today based on how young people use social media, as well as the recent ethical challenges facing military

37. *Marine Corps Strategy* 21, 2000; Major, A. E. "Ethics Education of Military Leaders." *Military Review* (2014 Mar-Apr): 55-60.; Bullis, C. and Reed, G. U.S Army War College. *Assessing Leaders to Establish and Maintain Positive Command Climate.* Department of Command, Leadership and Management Study, i-71. Carlisle, PA: Carlisle Barracks Printing, 2003.; Weber, J., and Gerde, V. "Organizational Role and Environmental Uncertainty as Influences on Ethical Climate in Military Units." *Journal of Business Ethics* 100 (2010): 595-612; Cullen, J., Victor, B., and Stephens, C. "An Ethical Weather Report: Assessing the Organization's Ethical Climate." *Organizational Dynamics* 18 no. 2 (1989): 50-62.
38. Keenan, J. "Protect What You've Earned: Engaging in the Conversation." *Marine Corps Gazette* 101 no. 3 (2017): 60-62.
39. Whiffen, H. "Becoming an Adaptive Leader." *Military Review*, Special Edition: The Profession of Arms (Sept 2007): p. 93.
40. Heifetz, R., and Linsky, M. Leadership on the Line: Staying Alive through the Dangers of Leading. Boston, Massachusetts: *Harvard Business Review* Press, 2002, p. 13.

commanders as technology continues to become more sophisticated. The Marine Corps' recent social media scandal illustrates the commander's operating environment and the need for commanders to understand how the young troops communicate.[41] These types of ethical challenges continue to plague military commanders.

Miller and Poole stress the importance of leaders being able to operate outside of their normal habitat, in unfamiliar environments, and with unfamiliar stakeholders—all of which enhance the complexity of the problem.[42] These rapid, dynamic changes associated with the shifts in societal attitudes coupled with the dynamics of the modern, complex battlefield point to the importance of sound ethical decisions.[43] General Martin E. Dempsey, former Chairman of the Joint Chiefs of Staff (CJCS), argues that senior military leaders need strong ethical fiber to assist them in dealing with the complex challenges associated with today's strategic environment.[44] Specifically, Dempsey relates military competence to personal ethics and the importance of dealing with ethical gray zones related to the complexities of the 21st Century security environment.[45]

Kahneman suggests that "intuition cannot be trusted in the absence of stable regularities in the environment."[46] Military leaders will experience tough problems in unstable environments,[47] as evidenced by the direction the United States Army provides its leaders on problem framing: identify complex issues, rely on their experience, expertise, and

41. Katolin, D. "Ethical Development: How We Shape Our People's Behavior." *Marine Corps Gazette* 100 no. 9 (2016a): 53-57.
42. Miller, C., and Poole, E. *Ethical Leadership, Global Challenges and Perspectives*. New York, NY: Palgrave MacMillan, 2011.
43. Wong, L. Developing Adaptive Leaders: The Crucible Experience of Operation Iraqi Freedom. Carlisle, PA, 2004. Whiffen, H. "Becoming an Adaptive Leader." *Military Review*, Special Edition: The Profession of Arms (Sept 2007): 92-98; Miller, C., and Poole, E. *Ethical Leadership, Global Challenges and Perspectives*. New York, NY: Palgrave MacMillan, 2011; Heifetz, R. Leadership Without Easy Answers. Cambridge, Massachusetts: Harvard University Press, 1994.
44. Major, E., DeRemer, L., and Bolgiano, D. "Ethics Can Be Taught." Proceedings 138 no. 12 (2012): 1-5.
45. Major, E., DeRemer, L., and Bolgiano, D. "Ethics Can Be Taught."; Major, A. E. "Ethics Education of Military Leaders." *Military Review* (2014 Mar-Apr): 55-60; Doty, J., and Doty, C. "Command Responsibility and Accountability." *Military Review* (2012 Jan-Feb): 35-38.
46. Kahneman, D. Thinking Fast and Slow. New York, NY: Farrar, Straus, and Giroux, 2011, p. 241.
47. DoD, US JFCOM, 2010, JOE.

judgment, and apply it to the context of the current situation.[48] Ethics education is valuable for all Services' senior leaders and is an integral component of a leader's ability to command military organizations as directed under Title 10.[49]

Ethics Education and Development

In 2012, then-Commandant of the Marine Corps, General James Amos, directed a Corps-wide ethical stand-down to address ethical issues in the Marine Corps such as sexual assault, suicide, and ethics-related failures in combat.[50] Further, in his 2013 White Letter 3-13, General Amos remarked: "I have a duty to not only ensure our warfighting readiness is maintained but to keep the moral and ethical health of our institution front and center."[51]

Numerous authors have suggested one possible way to accomplish what General Amos suggests is to continue to strengthen ethics education and development programs for our commanding officers and senior leaders.[52] According to Montor, the aim of ethics education, therefore, is what many refer to as "character development"—the creation of morally upright persons through the installation of certain key qualities or dispositions of character commonly known as "virtues."[53] For senior military leaders and commanders, this is accomplished at the senior service colleges.

48. Department of the Army, Training and Doctrine Command, Field Manual 22-100, Army Leadership, 1999.
49. Major, E., DeRemer, L., and Bolgiano, D. "Ethics Can Be Taught."; Freedman, D. *Corps Business: The 30 Management Principles of the U.S. Marines*. New York, NY: Harper Collins Publishers, 2000; U.S. Code, Title 10, 1947.
50. Rowell IV, G. Marine Corps Values-Based Ethics Training: A Recipe to Reduce Misconduct, Strategy Research Project, 1-29). United States Army War College. Carlisle, PA: Carlisle Barracks Printing, 2013.
51. Amos, J. "Command Climate.". White Letter No. 3-13. Headquarters Marine Corps, Washington, D.C., 2013a, p. 1.
52. Katolin, D. "Ethical Development: How We Shape Our People's Behavior." *Marine Corps Gazette* 100 no. 9 (2016a); Behn, B. U.S. Air War College. "The Stakes are High: Ethics Education at US War Colleges." Graduation Thesis, Air University, 1-27. Maxwell AFB, AL: Air University Printing, 2016.; Rowell IV, G. Marine Corps Values-Based Ethics Training: A Recipe to Reduce Misconduct, Strategy Research Project, 1-29); Major, A. E. "Ethics Education of Military Leaders." *Military Review* (2014 Mar-Apr): 55-60.
53. Montor, K. *Ethics for the Junior Officer*. Annapolis, MD: Naval Institute Press, 2001, p. 1.

The senior service colleges (or war colleges) teach ethics to military leaders, so they have a predisposition to ethical decision making.[54] For example, the Naval War College, Air War College, and the Army War College have an "ethics across the curriculum" approach where ethics instruction is included in a variety of disciplines sending a message to students that ethics plays a role in every aspect of their profession.[55] The Marine Corps War College (MCWAR) provides a stand-alone block of ethics instruction. Ethics is also included in other MCWAR curricula.[56] The Marine Corps and the Navy have a resident "ethics team" for teaching, curriculum, and staff development.[57] Further, the MCWAR program conducts "Director Calls" where the Director of the college meets with students to discuss strategic-level ethical issues and with general officers to discuss the application of ethics at the strategic level of war.[58] Relative to the war colleges, scholars and military authors argue that by providing ethics as a stand-alone course in the war college programs, students receive a clear message on the importance of ethics and its significance to commanders.[59] However, as discussed in the introduction, the military continues to experience ethical failures that must be addressed.

In 2012, then-CJCS, General Martin Dempsey addressed the faculty and student body at The National Defense University. He said: "For the first time, our competence and character are being evaluated by experts and pundits while we fight. ...There will be an ever-increasing expectation of service women and men to achieve that intricate balance of high moral character and high competence."[60]

54. Behn, B. U.S. Air War College. "The Stakes are High: Ethics Education at US War Colleges." Graduation Thesis, Air University, 1-27.; Rowell IV, G. Marine Corps Values-Based Ethics Training: A Recipe to Reduce Misconduct, Strategy Research Project, 1-29); Major, A. E. "Ethics Education of Military Leaders." *Military Review* (2014 Mar-Apr): 55-60; Major, E., DeRemer, L., and Bolgiano, D. "Ethics Can Be Taught." Proceedings 138 no. 12 (2012): 1-5.
55. Behn, B. U.S. Air War College. "The Stakes are High: Ethics Education at US War Colleges."
56. *Ibid.*
57. *Ibid.*
58. *Ibid.*
59. Rowell IV, G. Marine Corps Values-Based Ethics Training: A Recipe to Reduce Misconduct, Strategy Research Project, 1-29); Major, A. E. "Ethics Education of Military Leaders." *Military Review* (2014 Mar-Apr): 55-60; Major, E., DeRemer, L., and Bolgiano, D. "Ethics Can Be Taught." Proceedings 138 no. 12 (2012): 1-5; Wortel, E. and Bosch, J. "Strengthening Moral Competency: A Train the Trainer Course on Military Ethics." Journal of Military Ethics 10 no. 1 (2011): 17-35.
60. Major, A. E. "Ethics Education of Military Leaders." *Military Review* (2014 Mar-Apr): p. 58.

The military is governed by the Uniform Code of Military Justice, which parallels but does not replace, the laws in the U.S. Constitution.[61] Commanders are educated on the use of the Code and possess non-judicial punishment authority over their personnel enabling them to sanction individuals for minor infractions of the Code.[62] Marines have a code of ethics—honor, courage, and commitment—which governs the ethical conduct of all Marines.[63] These core values are taught to all Marine Corps personnel as part of ethics instruction. A consensus of scholars suggests a values- or virtues-based approach as the baseline for all ethics education in the military.[64]

One could argue that if the leaders are taught ethics differently than the followers it may cause confusion when it comes to reinforcing ethical behavior.[65] One critique of ethical theories comes from Robinson, who argues that some scholars question the utility of educating the rank-and-file on ethical philosophies that seem impractical or irrelevant to the military commander.[66] He adds that some military leaders debate the value of teaching the officers the philosophical underpinnings associated with the concepts, while other scholars argue the legitimate need for some instruction in ethical philosophy. The lack of standardization between the leaders and the subordinates could impact the leaders' ability to get the troops to think ethically and act autonomously.[67] This idea is significant, as Behn writes: "War college graduates play an essential role in establishing an ethical climate across the Joint Force and in maintaining trust between the military and the American public."[68] Therefore,

61. U.S. Code, Title 10, 1947.
62. *Ibid.*
63. Krulak, C. "Marine Corps Values Program." Marine Corps Order 1500.56. Headquarters Marine Corps. Washington: D.C. Government Printing, 1996b.
64. Robinson, P., De Lee, N., and Carrick, D. Ethics Education in the Military. Burlington, VT: Ashgate Publishing, 2008; Wortel and Bosch, 2011; Katolin, D. "Ethical Development: How We Shape Our People's Behavior." *Marine Corps Gazette* 100 no. 9 (2016a): 53-57.
65. Katolin, D. "Ethical Development: How We Shape Our People's Behavior." *Marine Corps Gazette* 100 no. 9 (2016a): 53-57; Major, E., DeRemer, L., and Bolgiano, D. "Ethics Can Be Taught." Proceedings 138 no. 12 (2012): 1-5.
66. Robinson, P. "Ethics Training and Development in the Military." *Parameters* 37 no. 1 (2007): 23-36.
67. *Ibid.*
68. Behn, B. U.S. Air War College. "The Stakes are High: Ethics Education at US War Colleges." Graduation Thesis, Air University, 1-27. Maxwell AFB, AL: Air University Printing, 2016, p. 18.

it is logical to assume that for the commanding officers to establish an ethical climate, their training and education should include an ethical, theoretical framework.

Theoretical Framework

The two types of ethical climate research that help define and frame the concept of ethical climate are the generic or molar approach and the focused or strategic approach.[69] While the molar field is concerned with capturing the overall sense of employees' experiences at work, the focused approach concentrates on a specific issue or outcome.[70] Research in molar climate examines and accounts for the entire meaning of the organization to the employees[71] The focused climate research field is concerned with the alignment of policies, practices, and procedures that focus on a specific organizational issue (e.g., ethics) or outcome (e.g., safety).[72] Kuenzi and Schminke[73] categorize focused climates into four categories: behavioral guidance, (e.g., justice, and ethics), involvement, (e.g., participation, and support), development, (e.g., innovation and creativity), and core operations (e.g., safety and service).[74] An ethical climate is a focused climate but is also linked to employee perceptions about "the alignment of various aspects of an organization's policies, procedures, and rewards that sends a consistent message to the

69. Erhart, M., Schneider, B., and Macey, W. *Organizational Climate and Culture: An Introduction to Theory, Research, and Practice*. New York, NY: Routledge, 2014.
70. *Ibid.*
71. *Ibid.*
72. Erhart, M., Schneider, B., and Macey, W. *Organizational Climate and Culture: An Introduction to Theory, Research, and Practice*; Kuenzi, M. and Schminke, M. "Assembling Fragments into a Lens: A Review, Critique, and Proposed Research Agenda for the Organizational Work Climate Literature." *Journal of Management* 35 no. 3 (2009): 634-717.
73. Kuenzi, M. and Schminke, M. "Assembling Fragments into a Lens: A Review, Critique, and Proposed Research Agenda for the Organizational Work Climate Literature." *Journal of Management* 35 no. 3 (2009): 634-717.
74. Erhart, M., Schneider, B., and Macey, W. *Organizational Climate and Culture: An Introduction to Theory, Research, and Practice.*

employees about what is valued in the organization."[75] Focused climates can be further divided into strategic climates and process climates.[76]

A climate is considered strategic when environmental influences drive a specific decisive outcome that can be tested by external standards, (e.g., a service or safety climate).[77] In contrast, a process climate focuses on "internal processes that occur in organizations as part of daily organizational functioning such as, a procedural justice climate, or an ethical climate."[78] Although scholars appear to treat each of the two types of climates—molar and focused—as separate,[79] process climates act as mediators in producing strategic outcomes. For example, process climates can provide a receptive foundation whereby employees feel the environment is ethical and fair, making them more amenable to management's strategic goals for employee effort and behavior.[80] Much of this work stems from the early efforts associated with ethical climate theory.

Victor and Cullen developed ethical climate theory (ECT), which is a type of climate that deals with organizational policies, practices, and procedures with moral consequences. [81,82] Ethical climate is a subset of

75. Erhart, M., Schneider, B., and Macey, W. *Organizational Climate and Culture: An Introduction to Theory, Research, and Practice*. New York, NY, pp. 85-90; Bowen, D. and Ostroff, C. "Understanding HRM-firm performance linkages: The Role of "strength" of the HRM system." Academy of Management Review 29 (2004): 203-221.
76. Schneider, B., Ehrhart, M., and Macey, W. "Organizational Climate Research: Achievements and the Road Ahead." In Handbook of Organizational Culture and Climate, 2nd edition, edited by N.M. Ashkanasay, C.P.M. Wilderom, and M.F. Peterson (29-49). Thousand Oaks, CA: Sage, 2011a, 2011b; Erhart, M., Schneider, B., and Macey, W. *Organizational Climate and Culture: An Introduction to Theory, Research, and Practice.*
77. Erhart, M., Schneider, B., and Macey, W. *Organizational Climate and Culture: An Introduction to Theory, Research, and Practice.*
78. *Ibid*, p. 87.
79. *Ibid.*
80. *Ibid.*
81. Simha, A., and Cullen, J. "Ethical Climates and Organizational Outcomes: Implications from the Past and Prophecies for the Future." *Academy of Management Perspectives* (Nov 2012): 20-34; Victor, B., & Cullen, J. B. (1987). A theory and measure of ethical climate in organizations. *Research in Corporate Social Performance and Policy* 9, 51-71; and Victor, B., & Cullen, J. B. (1988). The organizational bases of ethical work climates. *Administrative Science Quarterly* 33, 101-125.
82. Martin, K., and Cullen, J. "Continuities and Extensions of Ethical Climate Theory: A Meta-Analytic Review." *Journal of Business Ethics* 69 no. 2 (2006): 175-194.; Mayer, D., Kuenzi, M., and Greenbuam, R. "Examining the Link between Ethical Leadership and Employee Misconduct: The Mediating Role of Ethical Climate." *Journal of Business Ethics* 95 no. S1 (2010): 7-16.; Schminke, M., Arnaud, A., and Kuenzi, M. "The Power of Ethical Work Climates." *Organizational Dynamics* 36 no. 2 (2007): 171-186.

organizational climate and can be defined as "the perception of what constitutes right behavior, and thus becomes a psychological mechanism through which ethical issues are managed."[83] Ethical climate guides employees' ethical decision making and behavior.[84] ECT has both a sociological and ethical philosophical underpinning.[85]

Victor and Cullen include three dimensions of ethical philosophy in their theoretical strata: egoism, benevolence, and principle.[86] Egoism points to self-interest, benevolence—like utilitarianism—tries to produce the "greatest good for the greatest number of people," and principles like deontology guide decisions based on rules, codes, procedures, and laws.[87] These three areas constitute the ethical philosophy of the ECT framework. The sociological dimension of the theory consists of three loci: individual, local, and cosmopolitan.[88]

Simha and Cullen state: "The sociological component of the theory guides the individuals' decision making based on their own personal beliefs and values, the values of the organization, and the values of the

83. Martin, K., and Cullen, J. "Continuities and Extensions of Ethical Climate Theory: A Meta-Analytic Review." *Journal of Business Ethics* 69 no. 2 (2006), p. 177; Simha and Cullen, 2012, p. 21; Mayer, D., Kuenzi, M., and Greenbuam, R. "Examining the Link between Ethical Leadership and Employee Misconduct: The Mediating Role of Ethical Climate." *Journal of Business Ethics* 95 no. S1 (2010): p. 8-10.
84. Martin, K., and Cullen, J. "Continuities and Extensions of Ethical Climate Theory: A Meta-Analytic Review": 175-194.; Schminke, M., Arnaud, A., and Kuenzi, M. "The Power of Ethical Work Climates." *Organizational Dynamics* 36 no. 2 (2007): 171-186.
85. Victor, B., and Cullen, J. (1987, 1988). *A Theory and Measure of Ethical Climate in Organizations.*
86. *Ibid.*
87. Simha, A., and Cullen, J. "Ethical Climates and Organizational Outcomes: Implications from the Past and Prophecies for the Future." *Academy of Management Perspectives* (Nov 2012): 20-34; Martin, K., and Cullen, J. "Continuities and Extensions of Ethical Climate Theory: A Meta-Analytic Review." *Journal of Business Ethics* 69 no. 2 (2006): 175-194; Weber, J., and Gerde, V. "Organizational Role and Environmental Uncertainty as Influences on Ethical Climate in Military Units." *Journal of Business Ethics* 100 (2010): 595-612.
88. Simha, A., and Cullen, J. "Ethical Climates and Organizational Outcomes: Implications from the Past and Prophecies for the Future"; Martin, K., and Cullen, J. "Continuities and Extensions of Ethical Climate Theory: A Meta-Analytic Review." *Journal of Business Ethics* 69 no. 2 (2006): 175-194; Weber, J., and Gerde, V. "Organizational Role and Environmental Uncertainty as Influences on Ethical Climate in Military Units."; and Erhart, M., Schneider, B., and Macey, W. *Organizational Climate and Culture: An Introduction to Theory, Research, and Practice.* New York, NY, pp. 85-90.

society external to the organization."[89] The sociological component of the theory "parallels Lawrence Kohlberg's three sociomoral perspectives—preconventional (individual), conventional (organization or work-group), and postconventional (beyond society)."[90] The intersection of these theoretical dimensions (ethical philosophical, and sociological) produces nine different ethical climate types (self-interest, company profit, efficiency, friendship, team interest, social responsibility, personal morality, company rules and procedures, and laws and professional codes).[91] However, empirically, there are five types of ethical climates that occur most often: instrumental, caring, independence, law and code, and rules.[92] The data discussed in this book will assist leaders on how to promote benevolent and principled climates while discouraging egoistic and other negative climates.

To examine the effects of these types of ethical climates, Martin and Cullen conducted a meta-analysis of the ECT theoretical strata. Using the ethical climate questionnaire (ECQ) developed by Victor and Cullen[93] to measure perceptions of ethical climates, Martin and Cullen compiled a meta-analysis of 42 studies with numerous independent samples. Their study concluded that caring climates have a positive effect on employees and instrumental climates a negative impact on outcomes. Further, externally-based rules and codes were positively associated with ethical climate, but internally-imposed rules have a negative impact on outcomes. When rules climates are perceived, they act as effective control mechanisms but do not produce attachment to the organization.

89. *Ibid.*
90. Weber, J., and Gerde, V. "Organizational Role and Environmental Uncertainty as Influences on Ethical Climate in Military Units." *Journal of Business Ethics* 100 (2010): 595-612, p. 596.
91. Simha, A., and Cullen, J. "Ethical Climates and Organizational Outcomes: Implications from the Past and Prophecies for the Future." *Academy of Management Perspectives* (Nov 2012): 20-34; Martin, K., and Cullen, J. "Continuities and Extensions of Ethical Climate Theory: A Meta-Analytic Review." *Journal of Business Ethics* 69 no. 2 (2006): 175-194; Weber, J., and Gerde, V. "Organizational Role and Environmental Uncertainty as Influences on Ethical Climate in Military Units." *Journal of Business Ethics* 100 (2010): 595-612.
92. *Ibid.*
93. Victor, B., and Cullen, J. (1987, 1988). *A Theory and Measure of Ethical Climate in Organizations.*

However, when employees feel valued, they become more loyal and trustworthy to the organization.[94]

The preponderance of the evidence from studies examining traditional organizational outcomes (e.g., commitment to the organization and turnover intentions) has revealed that egoistic climates have negative outcomes while benevolent and principled climates have positive outcomes.[95] Other examples include rules climate, which could result in creativity constraints for employees, and benevolent climates that may prevent leaders from making hard choices.[96]

Using the ECQ, Weber and Gerde examined, over a one-year period, ten military workgroups with 218 participants from the U.S. Army and the U.S. Air Force.[97] The study explored the extent to which a military member's organizational environment influences the member's ethical decision making. They discovered that work groups with a higher percentage of risk of mission failure were more prone to instrumental climates. They also found the greater the perceived threat to the small group the more egoistical reasoning was employed and that the greater the level of perceived environmental uncertainty, the more likely the group had an instrumental climate.[98] Their research also mentioned that in a 2006-2007 survey administered to military personnel serving in a combat zone, 40% stated they would not report a colleague for committing a war crime. Further, 40% were reluctant to report someone for a war crime, and that unethical behavior is a significant concern of military leaders.[99] One finding

94. Martin, K., and Cullen, J. "Continuities and Extensions of Ethical Climate Theory: A Meta-Analytic Review." *Journal of Business Ethics* 69 no. 2 (2006): 175-194.
95. Simha, A., and Cullen, J. "Ethical Climates and Organizational Outcomes: Implications from the Past and Prophecies for the Future." *Academy of Management Perspectives* (Nov 2012): 20-34.
96. *Ibid.*
97. Weber, J., and Gerde, V. "Organizational Role and Environmental Uncertainty as Influences on Ethical Climate in Military Units." *Journal of Business Ethics* 100 (2010): 595-612.
98. *Ibid.*
99. *Ibid.*

remained constant from researchers on ethical climate—the importance of the leader in establishing the ethical climate.[100]

Definitions

Scholars appear to agree that process climates such as ethics have a significant impact on the workers' viewpoints on the overall wellbeing of the generic or molar climate.[101] This helps lay the groundwork for more strategic climates that become specific predictors of organizational outcomes, (i.e., ethical behavior).[102]

In the mid part of the last century, ethical work climate (EWC) was seen in two lights. EWC was defined as what is ethically correct in relation to moral obligations in organizations.[103] A second definition gained greater relevance in both organizational and military settings.[104] In their work, Victor and Cullen define ethical climate as "shared perceptions regarding organizational policies, practices, and procedures that emphasize ethical content."[105] This definition includes the individual, the organization,

100. Simha, A., and Cullen, J. "Ethical Climates and Organizational Outcomes: Implications from the Past and Prophecies for the Future." *Academy of Management Perspectives* (Nov 2012): 20-34; Martin, K., and Cullen, J. "Continuities and Extensions of Ethical Climate Theory: A Meta-Analytic Review." *Journal of Business Ethics* 69 no. 2 (2006): 175-194; Weber, J., and Gerde, V. "Organizational Role and Environmental Uncertainty as Influences on Ethical Climate in Military Units." *Journal of Business Ethics* 100 (2010): 595-612; Mayer, D., Kuenzi, M., and Greenbuam, R. "Examining the Link between Ethical Leadership and Employee Misconduct: The Mediating Role of Ethical Climate." *Journal of Business Ethics* 95 no. S1 (2010): 7-16.
101. Erhart, M., Schneider, B., and Macey, W. *Organizational Climate and Culture: An Introduction to Theory, Research, and Practice*. New York, NY: Routledge, 2014.
102. Erhart, M., Schneider, B., and Macey, W. *Organizational Climate and Culture: An Introduction to Theory, Research, and Practice*. New York, NY: Routledge, 2014; Schneider, B., Ehrhart, M., and Macey, W. "Organizational Climate Research: Achievements and the Road Ahead." In *Handbook of Organizational Culture and Climate*, 2nd edition, edited by N.M. Ashkanasay, C.P.M. Wilderom, and M.F. Peterson (29-49). Thousand Oaks, CA: Sage, 2011a; Kuenzi, M. and Schminke, M. "Assembling Fragments into a Lens: A Review, Critique, and Proposed Research Agenda for the Organizational Work Climate Literature." *Journal of Management* 35 no. 3 (2009): 634-717.
103. Victor, B., and Cullen, J. (1987, 1988). A Theory and Measure of Ethical Climate in Organizations.
104. Weber, J., and Gerde, V. "Organizational Role and Environmental Uncertainty as Influences on Ethical Climate in Military Units." *Journal of Business Ethics* 100 (2010): 595-612.
105. Victor, B., and Cullen, J. (1987, 1988). *A Theory and Measure of Ethical Climate in Organizations*, p. 101-125.

and the policies, practices, and procedures that commanding officers are sworn to carry out.[106]

Ethical climates set conditions for appropriate ethical behavior. Positive outcomes are created from ethical practices such as integrity, trust, justice, and social responsibility.[107] Schminke and colleagues suggest that ethical characteristics of the internal environment affect "how things are done around here."[108] Johnson states his definition: "Ethical climate, in turn, determines what members believe is right or wrong and shapes their ethical decision making and behavior."[109]

Weber and Gerde explore EWC in the military and define it as "a component of an organization's culture that influences a member's ability to recognize a problem, make a decision, and determine how to act appropriately."[110] Two decades earlier, Victor and Cullen defined EWC as "the perceived prescriptions, proscriptions, and permissions regarding moral obligations in organizations."[111] This added to their earlier definition, which focused on "the shared perceptions of what is ethically correct behavior and how ethical issues should be handled."[112]

Martin and Cullen define ethical climate as "the perception of what constitutes right behavior and thus becomes a psychological mechanism through which ethical issues are managed."[113] This definition is like that of Grojean, Resick, Dickson, and Smith, who argue that "the values

106. DON, HQMC, 1982, *The Marine Corps Manual.*
107. Barnes, C. and Doty, J. "What Does Contemporary Science Say About Ethical Leadership?" *Military Review* 90, no. 5 (2010): 90-93; Schminke, M., Arnaud, A., and Kuenzi, M. "The Power of Ethical Work Climates." *Organizational Dynamics* 36 no. 2 (2007): 171-186; Johnson, C. *Meeting the Ethical Challenges of Leadership, Casting Light or Shadow.* Thousand Oaks, CA: SAGE Publications, Inc, 2015.
108. Schminke, M., Arnaud, A., and Kuenzi, M. "The Power of Ethical Work Climates." *Organizational Dynamics* 36 no. 2 (2007): 171-186, p. 175.
109. Johnson, C. *Meeting the Ethical Challenges of Leadership, Casting Light or Shadow.* Thousand Oaks, CA: SAGE Publications, Inc, 2015, p. 321.
110. Weber, J., and Gerde, V. "Organizational Role and Environmental Uncertainty as Influences on Ethical Climate in Military Units." *Journal of Business Ethics* 100 (2010): p. 595.
111. Victor, B., and Cullen, J. (1987, 1988). *A Theory and Measure of Ethical Climate in Organizations,* p. 595.
112. *Ibid,* p. 52.
113. Martin, K., and Cullen, J. "Continuities and Extensions of Ethical Climate Theory: A Meta-Analytic Review." *Journal of Business Ethics* 69 no. 2 (2006): p. 177.

of the organization, its leaders, and its members play important roles in shaping the organization's climate regarding ethics."[114] In his White Letter 3-13, former Marine Corps Commandant, General Amos, wrote that "leaders at every level are responsible to create an environment and command climate in which every Marine is treated with dignity and respect; one which enables Marines to trust their command to...take appropriate action."[115] In other words, all leaders are responsible for reinforcing ethical behavior within the ranks.

Ethical Leadership

Johnson argues that various philosophers make the distinction between *ethics* as the study of right and wrong, and *morals,* which are described as standards of right and wrong. Both components are critical to ethical leadership, as the leader must make good choices and exhibit certain character traits like integrity, compassion, and justice to influence subordinates to act ethically.[116]

Military ethical leaders are professionals, and they place the greatest value on their service to society and adherence to professional ethics above compensation.[117] The military professional ethic (laws, values, beliefs) provides a set of standards that individual members must uphold to maintain the trust of the society they serve.[118] Ethical leadership is critical if commanding officers are to establish and maintain a command climate focused on ethical behavior.

Brown, Harrison, and Trevino define ethical leadership as "the demonstration of normatively appropriate conduct through personal action and interpersonal relationships, and the promotion of such conduct to

114. Grojean, M., Resick, C., Dickson, M., and Smith, D. "Leaders, Values, and Organizational Climate: Examining Leadership Strategies for Establishing an Organizational Climate Regarding Ethics." *Journal of Business Ethics* 55 no. 3 (2004): p. 226.
115. Amos, J. "Command Climate.". White Letter No. 3-13. Headquarters Marine Corps, Washington, D.C., 2013a, p. 2.
116. Johnson, C. *Meeting the Ethical Challenges of Leadership, Casting Light or Shadow.* Thousand Oaks, CA: SAGE Publications, Inc, 2015.
117. Lucas, G. *Routledge Handbook of Military Ethics.* New York, NY: Routledge, 2015.
118. *Ibid.*

followers through two-way communication, reinforcement, and decision making."[119] An ethical climate relies heavily on the influence from the commander. The two-way communication of what is perceived as ethically correct between the commander and the followers is critical. Schaubroeck and colleagues argue that ethical leaders at the highest levels have an ethical influence on the cognitions and behavior of lower-level leaders.[120] They illustrate this point through one of their studies.

Schaubroeck and colleagues were hired by the U.S. Army to conduct a study of soldiers deployed to Iraq in 2009, evaluating ethical conduct among other areas. Surveys were administered through Army chaplains and the Inspector General's office to more than 2,500 randomly selected soldiers. Survey results revealed that the effects of ethical conduct and shared ethical understandings at the highest levels influenced the leadership and shared ethical understandings at the lower-leadership levels.[121] Their study also examined various "embedding mechanisms"[122] discussed by Schein, such as paying attention to measuring and controlling followers' behavior, and the mechanisms of coaching and role modeling designed to instill ethical behavior in the followers.[123] Schein suggests other embedding mechanisms such as the allocation of rewards and status, resource allocation, and promotions as outcomes of ethical leadership.[124]

Haslam, Reicher, and Platow emphasize the moral aspect of ethical leadership.[125] They include in their definition of ethical leadership the

---

119. Brown, Harrison, and Trevino (2005) as cited in Schaubroeck, J., Hannah, S., Avolio, B., Kozlowski, S., Lord, R., Trevino, L., Dimotakis, N., and Peng, A. "Embedding Ethical Leadership Within and Across Organization Levels." *Academy of Management Journal* 55 no. 5 (2012): p. 1057.
120. Schaubroeck, J., Hannah, S., Avolio, B., Kozlowski, S., Lord, R., Trevino, L., Dimotakis, N., and Peng, A. "Embedding Ethical Leadership Within and Across Organization Levels." *Academy of Management Journal* 55 no. 5 (2012): 1053-1078.
121. *Ibid.*
122. Schein, E. *Organizational Culture and Leadership.* San Francisco, CA: Jossey-Bass, 2010.
123. Schaubroeck, J., Hannah, S., Avolio, B., Kozlowski, S., Lord, R., Trevino, L., Dimotakis, N., and Peng, A. "Embedding Ethical Leadership Within and Across Organization Levels." *Academy of Management Journal* 55 no. 5 (2012): 1053-1078; Schein, E. *Organizational Culture and Leadership.* San Francisco, CA: Jossey-Bass, 2010.
124. Schein, E. *Organizational Culture and Leadership.* San Francisco, CA: Jossey-Bass, 2010.
125. Haslam, S., Reicher, S., and Platow, M. *The New Psychology of Leadership: Identity, Influence, and Power.* New York, NY: Psychology Press, 2011.

importance of effectiveness and the value of orienting groups toward goals that are morally and socially responsible. In addition to these embedding mechanisms and emphasis on morals, ethical leadership requires integrity.[126] Ciulla and Forsyth explain that "a successful leader is someone who not only does the right thing but also does so in the right way and for the right reasons."[127] Further, Ciulla and Forsyth emphasize integrity, character, and ethical behavior as critical aspects of a leader's responsibility.[128] These traits and characteristics are important if leaders are to establish and maintain an ethical climate. Kirkpatrick and Lock studied key leadership traits or factors that leaders require to be effective, including honesty and integrity.[129] On a global scale, many researchers discuss the importance of integrity, values, and ethics to deal with stakeholders in an uncertain environment.[130]

Marine commanders should consider an ethical command climate to better accomplish their mission and to meet the expectations of the American people.[131] Although Katolin argues that the Marine Corps has work to do in ethics education,[132] and most would agree that it is the responsibility of the leader to ensure the subordinates are behaving in an ethical manner.[133] Katolin also argues that leaders must train and educate Marines on the value of ethics in all aspects of their profession.[134] Lenhardt adds that leaders' mentor their Marines continuously

126. Montor, K. *Ethics for the Junior Officer*. Annapolis, MD: Naval Institute Press, 2001.
127. Ciulla, J. and Forsyth, D. "Leadership Ethics." In *The Sage Handbook of Leadership*, edited by A. Bryman, D. Collinson, K. Grint, B. Jackson, and M. Uhl-Bien, 229-241. London: SAGE Publications, Ltd., 2011, p. 239.
128. *Ibid.*
129. Kirkpatrick, S. A., and Locke. E.A. "Leadership: Do traits matter?" *Academy of Management Executive* 5 no. 2 (1991): 48–60.
130. Miller, C., and Poole, E. *Ethical Leadership, Global Challenges and Perspectives*. New York, NY: Palgrave MacMillan, 2011; Pfeffer, J., Salancik, G., and Leblebici, H. *The Effect of Uncertainty on the Use of Social Influence in Organizational Decision Making*. Lexington, KY: The University Press of Kentucky, 1976; Heifetz, R. Leadership Without Easy Answers. Cambridge, Massachusetts: Harvard University Press, 1994.
131. Flynn, G. "Keepers of the Flame." *Joint Forces Quarterly*, 3d Quarter 54 (2009): 16-19.
132. Katolin, D. "Ethical Development: How We Shape Our People's Behavior." *Marine Corps Gazette* 100 no. 9 (2016a): 53-57.
133. Major, E., DeRemer, L., and Bolgiano, D. "Ethics Can Be Taught." *Proceedings* 138 no. 12 (2012): 1-5.
134. Katolin, D. "Military Ethics: How Maneuver Warfare Demands Adherence to Our Values." *Marine Corps Gazette* 100 no. 8 (2016b): 44-46.

by reinforcing ethical and moral essentials.[135] Also, military scholars suggest that reinforcing ethics education is critical to maintaining an ethical climate and is part of good (ethical) leadership.[136]

The Marine Corps conducted a survey of 82 Marine Corps junior officers in the Washington, D.C. area, who ranked ethical skills as the second most important skill set (behind interpersonal skills) as most pertinent to their ability to lead others.[137] Integral to ethical leadership is trust. As a group of Marine Corps lieutenants put it, "the American people deserve to know that they can always trust the Marines."[138] Therefore, it is logical to assume that ethical leaders must establish trust if they want to build unit cohesion.[139]

Katolin argues that ethics "is the foundation of facilitating trust."[140] Miller and Poole note that a leader must have integrity.[141] They cite Ciulla, who wrote that "the consensus of modern scholarship on leadership ethics is that integrity, character, and ethics are not just important, but they constitute the core of leadership responsibility."[142] Montor highlights the importance of trust as an integral part of ethics and argues that "the foundation for trust is being honest and telling the truth."[143] Johnson agrees that "leaders are key to the development of organizational trust."[144] Kirkpatrick and Lock's work illustrates key

135. Lenhardt, T. "Leading Marines: A Compassion Strengthened by Emotional Stability." *Marine Corps Gazette* 100 no. 6 (2016): 42-45.
136. Katolin, D. "Ethical Development: How We Shape Our People's Behavior." *Marine Corps Gazette* 100 no. 9 (2016a): 53-57; Lenhardt, T. "Leading Marines: A Compassion Strengthened by Emotional Stability." *Marine Corps Gazette* 100 no. 6 (2016): 42-45; Hipple, J., and Olson, S. "Values-Based Leadership: Lessons from the United States Marine Corps." *Harvard Business Review* (2010): 1-24.
137. Bailey, J. "Interpersonal Skill Development: A Requirement for TBS." *Marine Corps Gazette* 100, no. 7 (2016): 10-13.
138. (6th Platoon, Delta Company, The Basic School, 2016).
139. Katolin, D. "Military Ethics: How Maneuver Warfare Demands Adherence to Our Values." *Marine Corps Gazette* 100 no. 8 (2016b): 44-46.
140. Katolin, D. "Military Ethics: How Maneuver Warfare Demands Adherence to Our Values." *Marine Corps Gazette* 100 no. 8 (2016b): 44-46, p. 44.
141. Miller, C., and Poole, E. *Ethical Leadership, Global Challenges and Perspectives*. New York, NY: Palgrave MacMillan, 2011.
142. Ciulla (2011), cited in Miller, C., and Poole, E. *Ethical Leadership, Global Challenges and Perspectives*. New York, NY: Palgrave MacMillan, 2011, p. 209.
143. Montor, K. *Ethics for the Junior Officer*. Annapolis, MD: Naval Institute Press, 2001, p.7
144. Johnson, C. *Meeting the Ethical Challenges of Leadership, Casting Light or Shadow*. Thousand Oaks, CA: SAGE Publications, Inc, 2015, p. 337.

leadership traits or factors that leaders require to be effective, with one of these traits being honesty and integrity.[145] Other scholars discuss the importance of integrity, values, and ethics to deal with stakeholders and an uncertain environment.[146] Relative to trust, General Charles Krulak, 31st Commandant of the Marine Corps said that "leadership is fundamentally a reflection of an individual's values, education, training, and experience. It is above all else a product of character."[147] Montor argues the relevance of ethics to the military officer by stating that "when an officer acts unethically, it may cast doubt on the integrity of the officer corps and the military in general, raising doubts, perhaps serious doubts, about the quality of the U.S. military in the eyes of citizens, voters, and taxpayers who support it."[148]

The Role of the Commander

The commanding officer is responsible for establishing the right command climate based on non-negotiable values that guide the organization in everything it does.[149] Title 10, of the U.S. Code, section 5947 directs commanding officers to possess exemplary conduct and charges them to instill the same in their subordinates.[150] Scholars argue that the leader is responsible for developing subordinates mentally and physically, but also establishing policies and practices that create an ethical climate.[151] Katolin wrote, "Leaders must strive to train and educate

145. Kirkpatrick, S. A., and Locke. E.A. "Leadership: Do traits matter?" *Academy of Management Executive* 5 no. 2 (1991): 48–60.
146. Pfeffer, J., Salancik, G., and Leblebici, H. *The Effect of Uncertainty on the Use of Social Influence in Organizational Decision Making*. Lexington, KY: The University Press of Kentucky, 1976; Heifetz, R. Leadership Without Easy Answers. Cambridge, Massachusetts: Harvard University Press, 1994; Cameron, K. "Responsible Leadership as Virtuous Leadership." *Journal of Business Ethics* 98 no. S1 (2011): 25-35; Montor, K. *Ethics for the Junior Officer*.
147. Freedman, D. *Corps Business: The 30 Management Principles of the U.S. Marines*. New York, NY: Harper Collins Publishers, 2000, p. xiii.
148. Montor, K. *Ethics for the Junior Officer*, p. 3.
149. Doty, J., and Gelineau, J. "Command Climate." *Army Magazine*, July 2008, 22-24.
150. U.S. Code, Title 10, 1947.
151. Lenhardt, T. "Leading Marines: A Compassion Strengthened by Emotional Stability." *Marine Corps Gazette* 100 no. 6 (2016): 42-45; Grojean, M., Resick, C., Dickson, M., and Smith, D. "Leaders, Values, and Organizational Climate: Examining Leadership Strategies for Establishing an Organizational Climate Regarding Ethics." *Journal of Business Ethics* 55 no. 3 (2004): 223-241; Allen, C. "Ethics and Army Leadership: Climate Matters." *Parameters* 45 no. 1 (2015): 69-83.

Marines on the value of ethics in maneuver warfare and the importance of being an ethical warrior."[152] A commanding officers' ability to establish the right climate is traced back to his or her character, abilities, and actions, which include ethical behavior.[153] In summary, the commanding officer sets the tone for the ethical climate of the organization.

The climate refers to the environment of the organization, and it is more short-term in nature based on the network of personalities within the organization.[154] One example is the rotation policy of military commanding officers, who typically change position (command) every two years. This policy could potentially have negative implications for units trying to establish and maintain ethical climates based on the personalities and the priorities of the commanding officers.[155] Regardless, Marine commanders are responsible for training their personnel, establishing an ethical climate, and developing ethical subordinates.[156]

Doty and Gelineau state that "the commander can set a climate that fosters open and honest communication—both up and down the chain of command."[157] Open communication plays a significant role in trust among subordinates and in the perceptions of members about what constitutes good ethical behavior, which assists the commander in avoiding ethical "blind spots."[158]

Ethical Blind Spots

Even when commanders attempt to establish ethical climates and make the best decisions possible, ethical blind spots can interfere with ethical choices.[159] Bazerman and Tenbrunsel argue that leaders have ethical

152. Katolin, D. "Military Ethics: How Maneuver Warfare Demands Adherence to Our Values." *Marine Corps Gazette* 100 no. 8 (2016b), p. 46.
153. Department of the Army, Training and Doctrine Command, *Field Manual Army Leadership*, 1999.
154. *Ibid.*
155. B. Kerl personal interviews, Dec 2016.
156. Katolin, D. "Ethical Development: How We Shape Our People's Behavior." *Marine Corps Gazette* 100 no. 9 (2016a): 53-57; U.S. Code, Title 10, National Security Act, 1947.
157. Doty, J., and Gelineau, J. "Command Climate." *Army Magazine*, July 2008, p. 24.
158. B. Kerl personal interviews, Dec 2016.
159. Bazerman, M., and Tenbrunsel, A. *Blind Spots: Why We Fail to Do what's Right and What to Do about it*. Princeton, NJ: Princeton University Press, 2011.

blind spots and they need to overcome unconscious distortions that cause people to participate in or sanction behaviors they would normally condemn. Commanders must be aware of ethical blind spots which can be thought of as the gap between who we want to be and who we are.[160] Johnson argues that blind spots are due to our "own inherent goodness that may blind us to potential conflicts of interests that can undermine our objectivity and tempt us to make selfish choices."[161] Allen argues that blind spots can occur because of pressure to meet objectives and deadlines, careerism, protection of livelihood, organizations with low morale, and ignoring unethical acts.[162]

Ethical fading is a process that removes ethical dimensions from the decision making process.[163] Research in behavioral ethics reveals that people behave differently when confronted with an ethical dilemma and, at times, the "want self" dominates over the "should self." Here, the person's behavior is driven by desires and emotions, while ethical motivations and principles begin to fade, (i.e., ethical fading), and are in conflict at the point of decision.[164] This theory involves three perspectives: prediction (forecasting errors), decision time (ethical fading, visceral responses), and recollection (memory revisionism, shifting standards).[165] Commanders may increase the likelihood of establishing an ethical climate when they are aware of this phenomenon.

160. Bazerman, M., and Tenbrunsel, A. *Blind Spots: Why We Fail to Do what's Right and What to Do about it.*
161. Johnson, C. *Meeting the Ethical Challenges of Leadership, Casting Light or Shadow*, p. 196.
162. Allen, C. "Ethics and Army Leadership: Climate Matters." *Parameters* 45 no. 1 (2015): 69-83.
163. Bazerman, M., and Tenbrunsel, A. *Blind Spots: Why We Fail to Do what's Right and What to Do about it.*; Johnson, C. *Meeting the Ethical Challenges of Leadership, Casting Light or Shadow*; Schminke, M., Arnaud, A., and Kuenzi, M. "The Power of Ethical Work Climates." *Organizational Dynamics* 36 no. 2 (2007): 171-186.
164. Bazerman, M., and Tenbrunsel, A. *Blind Spots: Why We Fail to Do what's Right and What to Do about it.*
165. *Ibid.*

Ethical fading may be a key reason why people act differently than what they predicted. Ethics scholars believe that moral awareness prompts moral behavior. However, for some people at the time of decision, ethical fading occurs, and the ethical dimension of the decision fades away from the person's view.[166] One example of this is the space shuttle *Challenger*, where business concerns for a delayed or postponed launch dominated the ethical considerations relative to crew safety.[167] For commanders to account for ethical blind spots, scholars argue that ethics must be included in the decision making process.[168]

166. *Ibid.*
167. Allison, G., and Zelikow, P. *Essence of Decision, Explaining the Cuban Missile Crisis*. New York, NY: Addison-Wesley Publishers Inc., 1999.
168. Horn, W. "A Leader's Tool." *Marine Corps Gazette* 100 no. 9 (2016): 37-39; Johnson, C. *Meeting the Ethical Challenges of Leadership, Casting Light or Shadow*; Bazerman, M., and Tenbrunsel, A. *Blind Spots: Why We Fail to Do what's Right and What to Do about it*. Princeton, NJ: Princeton University Press, 2011.

Photo by: angelsimon.

CHAPTER TWO

Interviewing the Commandants

The research for this study was gathered primarily from personal interviews with eight CMC. The CMC were contacted by Headquarters Marine Corps (HQMC) once they received my letter requesting their support and participation.[169] These general officers were the units of analysis for this research project and provided the data that addressed the primary concern of the study: What role does command climate play in influencing the ethical behavior within a Marine Corps command? They provided concrete responses to the two research questions: 1) How do the CMC describe a command climate that encourages ethical behavior among Marines? and 2) How does a commanding officer develop a command climate that promotes ethical behavior?

Prior to conducting the CMC interviews, a round table discussion was initiated with two retired colonels with extensive command experience, one active-duty lieutenant colonel involved in the curriculum for new commanding officers, and a civilian leadership professor at the LLI in Quantico, Virginia, who specializes in ethics and command climate. The purpose of the round table discussion was to validate the semi-structured interview guide questions prior to interviewing the CMC. Participants were asked the same questions from the interview guide that was used with the CMC. Once the round table discussion was concluded, the participants agreed that the interview guide questions were appropriate to ask the CMC and relevant to the primary concern and the research questions.[170]

Each commentary from the Commandant interviews is presented as an individual case study. Chapter three presents the results of the cross-case

169. See Appendix B.
170. See Appendix C.

analysis that compared the responses from this chapter. In the individual cases, the narratives are followed by a summary of the key themes that were revealed through the interview. An analogous summary of the main themes follows the cross-case comparison in chapter three. The cases discussed in this chapter are presented chronologically (i.e., in the order of when each general was Commandant). The interviews differed in length, which had an impact on the volume of information provided by each case. Each contains its own unique themes, as a function of the individual Commandant's interests and concerns. The cross-case comparison in chapter three identifies and then examines four common categories that emerge from the individual interviews.[171]

This research project covers over 30 years of senior Marine Corps leadership experience. Each Commandant section begins with background information surveying various significant events that occurred during that Commandant's tenure, in order to provide historical context. This is followed by a detailed discussion covering the data collected from the personal interview. The official biographies of each Commandant are included in Appendix E of this book.

171. See Appendix D for coding information.

GENERAL ALFRED M. GRAY, JR., 29TH COMMANDANT (1987-1991)

Background

In late June 1987, General Alfred M. Gray, Jr., received the official battle colors and became the 29th Commandant of the Marine Corps at the Marine Barracks in Washington, D.C.[172] At the time, there were approximately 198,555 active-duty Marines.[173]

Commensurate with General Gray's emphasis on education and training, early in his tenure the Marine Corps established the Marine Corps Combat Development Command (MCCDC) at Quantico, Virginia, standardizing all training and doctrine throughout the Corps. Transitioning from the Vietnam era, and as part of General Gray's philosophy that every Marine is a rifleman, the Corps also developed the concept of basic warrior training for all Marines, to be implemented at the Marine Corps Recruit Depots (San Diego and Parris Island). The goal was to increase the combat effectiveness of all Marines, enabling them to serve at an optimum level as infantry in defensive or offensive combat operations.[174]

In 1989, the Marine Corps established a new professional military education program for officers, staff non-commissioned officers, and non-commissioned officers designed to provide both resident and non-resident career-enhancing instruction. Included in the curriculum was a professional reading program focused on instilling Marine Corps values and traits, sharpening analytical and reasoning skills, and cultivating a Marine's general knowledge and awareness of the principles established by America's Founding Fathers. Also, the Marine Corps University (MCU) was established under MCCDC, to oversee both resident and non-resident professional military education policies and programs.[175]

172. See Appendix E.
173. DON, HQMC, *Marine Corps Command Chronology, 1987-1991.*
174. *Command Chronology, 1987-1991.*
175. *Ibid.*

In addition to the above innovations in training and education, Marines continued to participate in a variety of operations, such as assisting with the Exxon Valdez oil spill in Alaska, protecting U.S. bases in Panama, providing humanitarian assistance in the aftermath of Hurricane Hugo in Charleston, South Carolina, and conducting disaster relief operations following the San Francisco Bay area earthquake.[176]

In 1990, Marines participated in operations such as the evacuation of U.S. embassy employees and American citizens from Monrovia, Liberia, during which 2,438 persons from 30 countries were evacuated, the Persian Gulf crisis in Kuwait following Iraqi President Saddam Hussein's invasion of neighboring Kuwait (*Operation Desert Shield*), and, in 1991, *Operation Desert Storm*, during which U.S. and allied forces defeated the Iraqi Army.[177]

While Marines were in Kuwait, the Corps was conducting other operations around the globe, including providing disaster relief after the eruption of Mt. Pinatubo in the Philippines, where approximately 6,000 Marines and Sailors provided support during *Operation Fiery Vigil*, and *Operation Provide Comfort*, during which Kurdish refugees in Turkey and Iraq received aid from Marines who established refugee camps and provided food and security to thousands of Kurds.[178]

Prior to General Gray's change of command, the Marine Corps began construction of the $14.2 million Marine Corps Research Center (Gray Research Center) in Quantico, VA, which would provide the Corps with a modern library, research, and conference facility to support professional military education. On June 28, 1991, General Gray presented the battle colors of the Marine Corps to General Carl E. Mundy, Jr., during a ceremony at the Marine Barracks in Washington, D.C.

---

176. *Ibid.*
177. *Ibid.*
178. *Ibid.*

The Interview

General Gray was interviewed at his office at the Potomac Institute for Policy Studies in Arlington, Virginia for approximately three hours. He emphasized that commanding officers should be leading Marines—not managing Marines—through a maneuver warfare philosophy in officers' approach to their command climate. He also stressed the importance of the commander setting the right example: "A leadership (not an academic) philosophy is defined as a system of ideas and a sum of ideals (personal convictions). A leadership philosophy is best conveyed through intent and example."[179]

This section discusses the commander's philosophy of command, which, according to General Gray, must build upon a foundation of trust and confidence, promote taking the initiative and responsibility, and include open and implicit communication. The section stresses the significance of setting an example, where the commander must take care of the people, demonstrate selflessness, be transparent, and ensure the Marines are having fun.

Philosophy of Command (A Maneuver Warfare Approach)

The commander's philosophy of command is instrumental in the command climate's effectiveness in promoting ethical behavior among Marines. The commander's philosophy of command includes "broad guidance in the form of concepts and values…a philosophy for action…a way of thinking in general."[180] General Gray discussed the maneuver warfare way of thinking and how it can be applied to command climate. Philosophy of command using a maneuver warfare approach entails boldness and initiative, competent leadership at every level, decentralized decision making while exercising sound judgment, implicit communication, and confidence among seniors and subordinates.[181]

179. Gray, A. and Otte, P. *The Conflicted Leader and Vantage Leadership*. Columbus, OH: Franklin University Press, 2006, p. 116..
180. *Ibid.*
181. Department of Defense, United States Marine Corps. *Warfighting*. New York, NY: Currency and Doubleday, Dell Publishing Group, Inc., 1989.

In the maneuver warfare approach to leadership, the commander decentralizes decision making to subordinate commanders down to the lowest level possible. This generates a tempo of operations that Marines desire "to best cope with the uncertainty, disorder, and fluidity of combat."[182] General Gray pointed out the importance of explaining to subordinates the rationale behind basing command decisions on ethical standards. He emphasized that "people need to know why they are doing what they're doing." The maneuver warfare approach can be followed both in combat and in peacetime. It must be built on trust and confidence, individual initiative and responsibility, and implicit communication.[183]

Commanders build trust and confidence by establishing long-term relationships with people. General Gray quoted the 13th Commandant of the Marine Corps, General John A. Lejeune's aphorism "You owe it to the young people to make them stronger physically, mentally, and morally." To establish trust with his subordinates, the commander must create an environment that facilitates creative thinking and does not punish individuals for taking the initiative or making mistakes. Even though the commander is still responsible for the Marines and the accomplishment of the mission, by creating a learning environment, the commander enables the Marines to think through problems and develop solutions on their own. As General Gray put it, "everyone kicks one in the grandstands once in a while."[184]

People make mistakes. Yet, through creating a learning environment of critical thinking, the commander builds the trust of the individuals in the command and builds their confidence by allowing them to make mistakes, as they attempt to follow the commander's intent. A corollary to this approach is the practice of reducing the number of rules and regulations, and keeping the rules simple.

---

182. Department of Defense, United States Marine Corps, *Warfighting*, 1989, p. 79.
183. Gray, A. and Otte, P. *The Conflicted Leader and Vantage Leadership*. Columbus, OH: Franklin University Press, 2006.
184. Otte, P. *Grayisms*, 2015, p. 92.

General Gray suggested that there is risk associated with too many complicated rules governing ethical behavior, which tend to become a "recipe climate" rather than a climate based on shared values, beliefs, and principles. He used the Army as an example of a large organization that relies on too many rules and procedures for everything. He suggested that this detracts from the individual initiative that is necessary for the sound judgment associated with good ethical decision making. In order to foster creative thinking and the confidence of Marines, the commander's philosophy must facilitate individual initiative and emphasize personal responsibility and accountability through good discipline.

Marines must have discipline in order to take the initiative and act responsibly. General Gray defined discipline by using an example from 1979, when he heard General Robert Barrow (27th Commandant) ask a Marine recruit in San Diego what discipline meant to him. The recruit responded "discipline means doing what must be done." This simple concept must be part of the command philosophy, according to General Gray, and reinforced by the commander on a regular basis. As General Gray put it, "you do what has to be done, and you do it the right way for the right reasons."

He mentioned that America expects its Marines to be special. The reason America has a Marine Corps, in his view, is that America wants a Marine Corps and believes Marines are good for the country. Referring to the "Marines United" scandal, General Gray commented that Marines have an obligation to be responsible and to behave as Marines 24/7, 365 days a year. In other words, "there's no such thing as off-duty." Social media, emails, text messages, etc., should be considered as signals security. If anything conveyed on social media could jeopardize the mission or cheat someone out of their dignity, then it is contradictory to good order and discipline and people who violate orders should be held accountable. For General Gray, the Golden Rule—treat people the way you would like to be treated—should underlie the behavior of Marines.[185] He mentioned that one of the things that makes Marines special is "the discipline, particularly the self-discipline of the Marine." In keeping with this attitude, General Gray made the issue of how male Marines were treating female Marines paramount shortly after becoming Commandant. General Gray wrote a letter to all commanders that stated "any Marine who doesn't treat my women Marines with dignity is out of step with their Commandant." General Gray spoke to roughly 9,000 of his 10,000 female Marines in 1987, and the consensus was that they just wanted to be treated with dignity.

Another key aspect of the commander's command philosophy and climate is implicit and open communication. The maneuver warfare approach requires implicit communication, through which the leader can communicate mutual understanding and a shared philosophy.[186] General Gray discussed the importance of good communication as a critical aspect of the commander's climate. He gave the example of his first day as Commandant at Headquarters Marine Corps. The workspaces, General Gray recalled, looked like a morgue with doors and windows closed. So, one of the first things he did was direct all personnel to open their doors. He started walking around talking to people to get to know them and allow them to understand his perspective as Commandant.

At the same time, General Gray emphasized open communication. In a good command climate Marines should be free to speak their minds, relying on the leadership of the command to listen to their thoughts, ideas, and recommendations. General Gray pointed out that "there are times when you want to leave your rank behind, particularly at critiques and after exercises." In other words, critiques should be conducted by small unit leaders and not driven by senior commanders who do

185. *Ibid*, p. 93.
186. Gray, A. and Otte, P. *The Conflicted Leader and Vantage Leadership*, 2006.

all the talking. To foster both implicit and open communication, the commander must be a good listener and be able to sense what is going on within the command. This is part of the commander's responsibility to constantly evaluate his or her climate and determine if Marines are making good ethical decisions.

Setting the Example (Living it)

General Gray discussed the importance of setting the example: "The commander should not only set it but must live it." As a commanding officer, you must "walk your talk."[187] General Gray outlined four critical areas instrumental in setting the example and establishing the right climate to properly influence the ethical behavior of Marines: 1) taking care of your people, 2) being selfless, 3) being transparent, and 4) having fun in your profession.

One of the most important things a commander does is to take care of followers through training and education. General Gray alluded to the 13th Commandant, General John A. Lejeune's message[188] regarding an officer's responsibility to the troops: "Make them better morally, ethically, and physically when they leave you than when you joined them." General Gray set a superior example of this aspect of leadership when he served as commander at Camp Hansen in Okinawa in the 1970s. During this period there were many racial problems, and approximately 55% of General Gray's Marines did not have their high school diplomas. So, he organized a group of officer's wives who had teaching credentials and formed a night school for his Marines to earn high school diplomas. This fostered teamwork and reduced the racial tensions within his organization. This example also showed the Marines that their commander believed in education and that he cared

187. *Ibid*, p. 46.
188. Lejeune, John A. *Reminiscences of a Marine*. University of Wisconsin-Madison, WI: Arno Press, 1930.

about them enough to help them further their education. As General Gray said during the interview, "you've got to take care of people, and that's part of your command environment."

As part of setting the example, General Gray stressed the significance of walking around and asking people how they were doing. In Okinawa, he would walk into the town of Kinville, which he knew was dangerous due to the racial tensions and the amount of alcohol Marines consumed while on liberty (off-duty). General Gray spoke with the owners of the bars and to the Marines about how he could help reduce the number of disorderly conduct incidents happening out in town. Eventually, he established a curfew and courtesy patrols to ensure Marines were behaving appropriately and returning safely to the barracks.

Another instance was that General Gray did not like the fact that many of the officers living in town were showing up late for work and setting a bad example. So, he ordered the officers to move on to Camp Hansen and live on the base like their troops who lived in the barracks. Again, this gesture demonstrated how much Gray believed that leaders must set the example and that commanders must be focused on their troops and the Marine Corps, not themselves.

A second aspect of setting the example is selflessness. General Gray pointed out that "you can get anything done you want if you don't care who gets the credit for it." He inherited this principle from his father, and it was always paramount in his command climate. General Gray believes that when people know the leader is more concerned about them than himself, they will have a greater tendency to exhibit better behavior. He mentioned that his "greatest legacy" and source of happiness was the people he developed throughout his career that became great leaders, such as General Jack Sheehan and General Anthony Zinni: "Your greatest success is when the young people that you're privileged to teach get to be better than you are. So, I think that's all part of your climate." He went on to say that "unless you care more about others than yourself, you will fall prey to careerism."[189]

General Gray would talk to the families of the Marines to ensure they were doing well while their spouses were away. As General Gray once remarked to General Anthony Zinni, "we now have more dependents in the Marine Corps than we have Marines. So, your challenge then has expanded." General Gray emphasized that commanders must take care of their Marines, as well as their families, to set the right climate for good ethical behavior.

A third aspect of setting an example is transparency. When General Gray became Commandant of the Marine Corps, he had a meeting with his Public Affairs Officer. The office told General Gray that the mission of Public Affairs was to ensure that the Marine Corps looked good. General Gray retorted that the mission of Public Affairs had just changed. He decreed that from that moment on "the mission of Public Affairs is to get the facts before the American people." The commander must be transparent whenever a mistake is made, and that must be part of the commander's climate. It is the right example for the commander to set.

Finally, General Gray discussed the importance of the fourth aspect of setting an example: the commander creating an environment where the Marines have fun and can enjoy their profession. When General Gray was at Camp Lejeune after a major exercise in the field, he received a phone call from the Secretary of the Navy, James Webb, who told him President Reagan had appointed him to be the next Commandant. General Gray responded, "Mr. Secretary, we're going to have some fun." General Gray highlighted the significance of intramural sports for building camaraderie and teamwork, as the kind of fun that is healthy in a command climate that builds teamwork and promotes unit cohesion. He suggested that commanders should focus on the kind of fun that

189. Otte, P. *Grayisms*, 2015, p. 52.

"makes people appreciate one another" as members of an elite team of professionals who enjoy going to work and being United States Marines.

Summary

General Gray argued that the philosophy of command replicates the Marines' doctrine of maneuver warfare, which is built on trust and confidence of subordinates, initiative, and an implicit understanding of what the commander wants to accomplish based on the close personal relationship between the commander and the subordinates. He emphasized the importance of setting the example by taking care of the people, being more concerned about the organization and the people than yourself, making sure the commander is completely transparent in his or her actions, and that the people have fun being United States Marines.

General Alfred Gray and the author.

GENERAL CHARLES C. KRULAK, 31ST COMMANDANT (1995-1999)

Background

On June 30, 1995, General Charles C. Krulak became the 31st Commandant of the Marine Corps in a change of command ceremony at Marine Barracks, Washington, D.C.[190] There were approximately 175,000 active-duty Marines in the Corps at the time.

As General Krulak became Commandant, Marines were involved in the North Atlantic Treaty Organization (NATO)-directed air strikes against Bosnian Serb military targets as part of *Operation Deliberate Force*. In addition, Marines participated in numerous exercises throughout the world, including exercises in Kuwait and Egypt. During the latter, 33,000 Egyptian, British, French, and Arab Emirates troops participated in the largest joint military exercise ever held in that region. To support exercises like these, General Krulak established the Commandant's Warfighting Laboratory (CWL) to serve as the test bed for the development of new concepts, tactics, techniques, procedures, and doctrine.[191] In 1995, General Krulak directed the relocation of Headquarters Marine Corps (HQMC) to the Pentagon adjacent to the office of the Secretary of the Navy. Further, he published the Commandant's Planning Guidance, which focused on the concept of "Transformation." Transformation was General Krulak's term for the process that takes citizens, transforms them into United States Marines for the 21st Century, and then returns them back to society better than before they came into the Corps.[192]

Other significant operations during this period included the conclusion of the Marine Corps' involvement in Haiti as part of the United Nations' peacekeeping mission, as well as Marine participation in providing security assistance to the U.S. Embassy compound in Monrovia, Liberia.[193]

190. See Appendix E.
191. Department of the Navy, HQMC, *Marine Corps Command Chronology, 1995-1999.*
192. Department of the Navy, HQMC, *31st Commandant's Planning Guidance*, 1995.
193. *Command Chronology*, 1995-1999.

Over the next two years, General Krulak made other changes, including a new approach in training Marines for warfighting with the introduction of the "Crucible," a 54-hour event at boot camp where recruits' mental, moral, and physical training experience culminated and transformed the civilian into a Marine. As part of this new warfighting spirit, all Marines were issued a plastic card with the Marine Corps' core values of honor, courage, and commitment, which they were required to sign, signifying their commitment to these values.[194]

In 1997, General Krulak implemented several other initiatives, including the CWL experiment Hunter Warrior, which involved 7,000 Marines and Sailors in Southern California testing future concepts and tactics, the introduction of female Marines shooting live ammunition from heavy weapons in combat training at Camp Lejeune, and the relocation of the 3d Marine Aircraft Wing to Marine Corps Air Station Miramar.[195]

Marines continued to provide a show of force in the Persian Gulf, while others were sent to Mombasa, Kenya to provide humanitarian assistance due to flooding in the northeastern part of the country. Concurrently, Marines supported NATO operations in Bosnia as a ready reserve force and conducted training in areas such as Thailand and near the southern coast of the Turkish Republic.

Many other global exercises and operations occurred until General Krulak relinquished his duties. These included Marines providing humanitarian assistance and disaster relief operations in response to Hurricane Mitch in Central America, Marine Corps air strikes against Iraq, assisting with the evacuation operation of approximately 90 American citizens from the U.S. embassy in Kuwait, and security and humanitarian assistance missions in Albania while supporting the air campaign against the Serb-led Yugoslavian government. On June 30, 1999, General James Jones replaced General Krulak, becoming the

194. *Ibid.*
195. *Ibid.*

32nd Commandant of the Marine Corps during a ceremony at Marine Barracks, 8th and I Streets, Washington, D.C.[196]

The Interview

General Krulak was interviewed at his home in Birmingham, Alabama for approximately three hours. The general began the interview by embracing the concern of the study: "Let me start by saying that, one, your overarching question is a critical one. Probably the most critical question facing the Corps today." General Krulak's interview centered on the importance of commanders continuing to develop the character of their Marines (sustaining the Transformation) and holding Marines to high, almost spiritual, standards of conduct.[197]

He established that the primary purpose of a command climate is to influence the ethical behavior of the Marines in the unit. The climate of the Marine Corps begins with the Commandant, stated General Krulak, and descends down to every commanding officer in the Corps. To highlight the significance of maintaining high, almost spiritual, standards, he read a quote from his late father, Lieutenant General Victor Krulak, USMC (Ret) who said:

> We exist today, and we flourish today, not because of what we know we are, or what we know we can do, but because of what the grassroots of our country believes we are and believes we can do. The American people believe that Marines are downright good for the country; that Marines are masters of a form of unfailing alchemy which converts unoriented youths into proud, self-reliant, stable citizens; citizens in whose hands the nation's affairs may safely be entrusted. And, likewise, should the people ever lose that conviction as a result of our failure to

196. *Ibid.*
197. Krulak, C. "Character." *Leatherneck Magazine*, 79 no. 12 (1996a): 14-15.

> meet their high, almost spiritual standards, the Marine Corps will then quickly disappear.[198]

General Krulak's Commandant's Planning Guidance (CPG) was the primary instrument he used to establish command climate for the Marine Corps. Through it, he promoted his two key initiatives: The Transformation and the Marine Corps Values Program. This section of the book addresses these two initiatives, which were emphasized by General Krulak during his interview.

The 31st Commandant's Planning Guidance

According to General Krulak, the two most important things the Marine Corps does for the nation are win battles and make Marines. To continue to achieve these two things, the CPG was developed by General Krulak to outline the Corps' strategic approach on "where the Marine Corps is going and why, what the Marine Corps will do, and in some instances, how and when prescribed actions are to be implemented."[199]

The CPG emerged from General Krulaks' most strongly held beliefs as five pillars: 1) warfighting, 2) people, 3) core values, 4) education and training, and 5) the Marine Corps' naval character.[200] Two of these five pillars are most relevant to the concerns of this study: education and training and the core values.

During the interview, General Krulak suggested that we need to get back to educating our Marines on what it means to be a Marine. Commanders must read and understand Title 10 of the U.S. Code and what the 82nd Congress signed into law regarding the Marine Corps' purpose. Initially, this was the result of a small group of Marines and politicians who, after World War II, worked to prepare a speech (the "bended knee" speech) for General Vandegrift, then-Commandant of the Marine Corps.

198. Krulak, V. *First to Fight: An Inside View of the U.S. Marine Corps.* Annapolis, MD: Naval Institute Press, 1984, p. xv.
199. DON, HQMC, The *31st Commandant's Planning Guidance*, 1995, p. A-1.
200. *Ibid.*

The speech was written to convince the 82d Congress and President Truman that the Marines should not be subsumed under the Army and the Air Force. The authors were a small group of Marines along with others who were nicknamed the "Chowder Society." After Vandegrift's speech, Congress was so impressed that they wrote into the law the basic principles for the organization of the Marine Corps which were part of the 1947 National Security Act.[201]

General Krulak discussed how commanders must constantly reinforce what it means to be a Marine. He believes that if Marines can commit to making good choices a daily habit, based on their training and education, they will have a greater tendency to draw from this habit when performing their duties in combat while under extreme stress. In other words, Marines must be ready to discharge the other duties as the president may direct, but, most of all, be prepared to fight and win as ethical warriors.

Being an ethical warrior, declared General Krulak, "is about the beliefs shared by all Marines, past, and present, that there's no higher calling than that of a Marine, and that being a Marine is more than excellence in combat. It is also excellence in character." As we shall see, General Krulak holds commanders responsible for developing their Marines' character based on the Marine Corps' core values, and for the sustainment of those values.

General Krulak's reliance on core values was discussed throughout the CPG. In the section on the future direction of the Marine Corps, General Krulak mentioned core values: "Our Marines' moral character, courage, and ethical values will dominate any location or operational area with the unconditional certainty that the Marine Corps is a force for good."[202] General Krulak added that "I do not intend for 'honor, courage, and commitment' to be just words. I expect them to frame the way we

201. U.S. Code, Title 10, National Security Act, 1947.
202. *31st Commandant's Planning Guidance*, 1995, p. A-3.

live and act as Marines."[203] Under General Krulak's direction, every facet of the Marine Corps emphasized the core values. This included how Marines were recruited, trained, employed, and prepared to eventually return to society after successfully completing their military service. This process was called the "Transformation."

The Transformation

The Transformation involved recruiting, recruit training, cohesion, and sustainment. It was designed to take a person from the civilian sector, recruit them, train them, develop them into Marines, and eventually return them back to society as people of sound character who live by the core values of honor, courage, and commitment.[204] The Transformation was developed based on the view of the world in 1995 that the 21st Century would find Marines fighting non-state actors (e.g., Al-Qaeda) using information technology (e.g., the CNN effect) coupled with terrorist tactics or asymmetric (non-traditional style) warfare. Marines would be involved in conflicts that required them to conduct humanitarian assistance operations in the morning, peacekeeping missions in the afternoon, and combat operations in the middle of the night. These missions could take place within three city blocks spanning a 24-hour period ("the three-block war"). Therefore, America needed a force with people who had the right values, and who possessed the moral strength of character to make good ethical decisions. Poor ethical decisions could contribute to the CNN effect by worldwide portrayal of negative images that contradict American values.

General Krulak called upon psychologists and psychiatrists to learn more about the young people coming into the Marines (at the time, Generation X). He reoriented the Corps' recruiting efforts to accommodate the customers—America's youth. These young people wanted to know the boundaries of acceptable behavior and would be good followers if they had the opportunity to become leaders. They also wanted

203. *Ibid*, p. A-17.
204. DON, HQMC, MCRP 6-11D, *Sustaining the Transformation*, 2014.

to belong to something that had value. The Marine Corps changed its recruiting tactics by raising its entrance standards to get more high school graduates who were right-thinking young men and women who wanted to become part of something greater than themselves. They wanted to be easily recognizable, believed in doing what is right, and aspired to follow a higher calling. These were the young people the Marine recruiters were sending to boot camp.

Recruit training was also changed as part of the Transformation, by becoming more rigorous and challenging. General Krulak directed an increase in the total number of hours at recruit training and implemented the following changes: he made the physical fitness test the same for both men and women, gave more time back to the drill instructors to mentor recruits on core values, and added the Crucible to recruit training.[205]

The Crucible was a 54-hour, mentally and physically grueling culmination of 11 weeks of recruit training. It reinforced instruction on teamwork, leadership, and the Marine Corps' core values of honor, courage, and commitment. Based on the changing world of technology coupled with asymmetric combat in an urban environment, General Krulak said of the Crucible that, "Marines must be good decision makers. They must be trained to the highest standard." In addition to rigorous training and high standards, for Marines to meet the challenges and demands of the current threats, Marine units needed to coalesce to realize their maximum potential as a warfighting organization.

General Krulak ordered his staff to promote more unit cohesion by "forming Marines into military occupational specialty teams, sending those teams to their occupational schools and then on to their first unit until the end of their initial enlistment."[206]

---

205. Krulak, C. "Sustaining the Transformation." White Letter No. 3-98. (1998).
206. *Ibid.*

Another example of how General Krulak promoted better unit cohesion within the Corps was with the female drill instructors. Before General Krulak became Commandant, female drill instructors wore a different cover (hat) when working with recruits and were not authorized to wear the prized drill instructor (DI) campaign cover (i.e., the "Smokey the Bear hat") that is worn by male instructors. General Krulak directed that all drill instructors wear the campaign hat. He did this at the same time he directed that all Marines regardless of gender would go through the Crucible during entry-level training. This was another move aimed to foster greater unit cohesion in the Marine Corps. Commensurate with General Krulak's guidance from the CPG, the responsibility of maintaining unit cohesion and sustaining the Transformation fell on the shoulders of the commanding officers.

General Krulak pointed out that "as commanders and leaders we are responsible for the character development of the Marines we lead. This phase in the Transformation process is where you, the leaders of our Corps, take over."[207] He referenced Marine Corps history when he recalled that "as commanders and leaders, we all share responsibility for ensuring that Marines embody those values we have cherished for more than two centuries."[208] He emphasized that Marines must uphold the legacy and the definition of being a Marine when he quoted the 13th Commandant, General John A. Lejeune, who said "[Marines are] the descendant of a line of heroes, the bearer of a name hailed as foremost in the annals of his country, the custodian of a long-cherished reputation for honor, valor, and integrity."[209]

During the interview, when General Krulak referred to the primary concern of the study, he said, "I really never strayed from your central premise which was: what role does the command climate play in influencing the ethical behavior of the Marine Corps. What a critical discussion of what's going to keep the Marine Corps alive." This emphasizes the significance of the command climate and the commander's role in reinforcing

207. Krulak, C. "Marine Corps Values Program." White Letter No. 16-96 (1996c).
208. *Ibid.*
209. *Ibid.*

the core values on a routine basis. General Krulak believes that this process starts with the Commandant and his commanding officers, who he charged directly responsible for sustaining the Transformation and maintaining character development programs within their units: "it's got to come from the top." In other words, the commanders must establish a climate that influences the ethical behavior of the Marines in their organizations.

Clearly, General Krulak feels strongly that in addition to fighting and winning the nation's battles, the most important thing the Marine Corps can do as an institution is to take young men and women from society and make them Marines of good character. And that the Marine Corps will eventually return these Marines back to their families, states, cities, and towns, as better people than before they joined the Corps.[210] It is the responsibility of the commanding officer to ensure that the Marine Corps' values are implemented and reinforced. The Marine Corps Values Program was designed by General Krulak to assist commanders with accomplishing that task.

Marine Corps Values Program

In the 1970s, the Marine Corps worked to improve its post-Vietnam image and the issue of drug abuse. During the 1980s and early 1990s, the Corps emphasized professional military education with its professional reading program and the establishment of the MCU. General Krulak emphasized "improving and enhancing every Marine's devotion to the values of honor, courage, and commitment which have been the hallmark of Marines since the founding of our nation."[211]

210. Krulak, C. *Implementing Instructions for the Marine Corps Values Program*. ALMAR 439/96, 1996d.
211. *Ibid*, p 2.

In addition to publishing his planning guidance, he directed all Marines to read *Rifleman Dodd,*[212] which is a book about the British Army fighting the French during the Peninsular Wars. General Krulak believed this book would help to further Marines' understanding of commitment to the core values.[213] In the book, after months of combat and little to eat, a private (Dodd) in the Ninety-Fifth Foot, a British infantry company, who was near starvation after months of combat, survived the ordeal and was happy to have done his duty in support of his fellow soldiers. Dodd cherished the fact that he was a member of the Ninety-Fifth "whose boast was that they were always first into action and last out."[214] General Krulak made commitment to the core values the theme for the Marine Corps in 1995.[215]

In December of 1996, General Krulak directed all commanding officers to use the tools that HQMC provided to assist them in sustaining the Marine Corps' values of honor, courage, and commitment: "Using the foundation set in entry-level training, take these tools, your creativity, and knowledge of your unit to craft a program that addresses the unique issues you face."[216] The Commandant's intent was for all commanders to ensure their command climates emphasized these enduring values and that they were reinforced on a regular basis.

Moreover, Marine Corps Order (MCO) 1500.56 directed that commanding officers "integrate Marine Corps values training into organizational training plans."[217] The tools mentioned in both the White Letter 16-96 and the MCO 1500.56 included a discussion guide for leaders, a Marine Corps values pocket card to be issued and signed by all Marines, and a Commandant video about the program.[218] General Krulak said in the message that he sent to all Marines: "Our core values remain the very

212. Forester, C.S. *Rifleman Dodd.* Baltimore, MD: The Nautical and Aviation Publishing Company of America, 1989.
213. Krulak, C. *Professional Reading Program; Commandant's Choice for 1995.* ALMAR 157/95, 1995.
214. (Forester, 1989, insert)
215. Krulak, C. *Professional Reading Program; Commandant's Choice for 1995.* ALMAR 157/95, 1995.
216. Krulak, C. "Marine Corps Values Program." White Letter No. 16-96, 1996c, p 1.
217. Krulak, C. "Marine Corps Values Program." Marine Corps Order 1500.56. Headquarters Marine Corps. Washington: D.C. Government Printing, 1996b, p. 3.
218. *Ibid.*

soul of our institution, underlying all that is best in Marines and must continue to frame the way we live and act as Marines."[219]

One issue discussed during the interview with General Krulak was the competing demands faced by commanders, which can potentially have a negative impact on the implementation and sustainment of the values program. Areas such as maintenance of equipment, mission essential task list training, and other required training, all compete for "white space" on a commander's annual training plan. On this topic, General Krulak opined that "you cannot excuse this away because of operational tempo or because we are at war, or any other number of things. It's got to be constantly at the forefront." Regarding the other tasks levied upon a commander, and highlighting the importance of the values program, General Krulak offered that "it begets warfighting. This begets excellence. I worry when we talk about balanced excellence. It's only balanced when the ethical part of the equation is first. It better be the first number in the equation."

General Krulak discussed the value of getting it right when it comes to the commander's climate and its influence on ethical behavior. He suggested that "if we get this right, the rest of it will come along. And you'll get more money from Congress, and you'll get more people to come into the Marine Corps."

Prior to, and following, the implementation of the Marine Corps Values Program, General Krulak published a series of messages to all Marines that discussed traits associated with a Marine's character. These messages were designed to help guide Marines in their ethical behavior, but also to assist commanders in establishing the right climate influencing the ethical behavior of the Marines in the unit. During his first two years as Commandant, General Krulak repeatedly emphasized the core values and made numerous changes to embed these values into everything the Marine Corps touched.

219. Krulak, C. *Implementing Instructions for the Marine Corps Values Program*. ALMAR 439/96, 1996d, p. 1.

In July of 1997, General Krulak published an update on the progress of the Marine Corps Values Program. This update offered insight as to how the program influenced the ethical behavior of Marines within the various units.[220] General Krulak discussed the progress made at the time using examples such as the staff at the Basic School for Marine lieutenants, who visited the Vietnam Memorial and the Marine Corps War Memorial to reflect on what Marines did in the past. After the discussion and reflection period, Marines were awarded their values cards in the shadow of the Marines raising the American flag at the top of Mount Suribachi on Iwo Jima.[221]

General Krulak's efforts as Commandant focused on a commitment to the Marine Corps' core values. As General Krulak said to his commanders: "This phase is the most difficult, and I purposely kept it non-prescriptive. I want you, the leaders of our Corps, to be aware of, and use the tools available to you in these efforts. Use your initiative, intelligence, and imagination to ensure the success of this program."[222] Before the interview concluded, General Krulak read the quote from his father again, to put added emphasis on the importance of what "Marine" stands for in the eyes of the American public, and why it is so important to the Corps' existence for all Marines to live up to those high, almost spiritual, standards.

Summary

General Krulak published his *31st Commandant's Planning Guidance*, which laid out his guidance on command philosophy and his commander's intent to the Marine Corps. It emphasized education and training, and a re-focus on the Marine Corps' core values of honor, courage, and commitment. General Krulak implemented changes that would promote greater unit cohesion within the Marine Corps, such as the Crucible

220. Krulak, C. *Report on the Progress of the Marine Corps Values Program*. ALMAR 238/97, 1997.
221. Krulak, C. *Report on the Progress of the Marine Corps Values Program*. ALMAR 238/97, 1997.
222. *Ibid*, p. 2.

at recruit training and a focus on sustaining a Marine's Transformation after the completion of entry-level training through the Marine Corps Values Program.

General Krulak held commanding officers directly responsible for sustaining the Transformation of Marines through their continual character development as the focal point of what it means to be a Marine. To assist the commanders in their duties, he published a series of messages to all Marines that highlighted various aspects of character such as integrity and courage. General Krulak's thesis was that, for the Corps to remain in existence, it is critical for all Marines to abide by the core values and to ensure the Corps never loses the trust of the American people, which is built on their sentimental ideas of what a Marine can do and what a Marine represents. Marines may not necessarily be needed for the survival of the nation, but Americans want Marines because they are good for the country.

General Charles Krulak and the author.

GENERAL JAMES L. JONES, 32ND COMMANDANT (1999-2003)

Background

In June 1999, General James L. Jones became the 32nd Commandant of the Marine Corps during a ceremony at Marine Barracks, 8th and I Streets, Washington, D.C., replacing General Charles C. Krulak.[223] At that time, there were approximately 172,000 active-duty Marines in the Corps. During this period, Marines were involved in operations at home and abroad, such as the relocation of Marine Corps Air Station (MCAS) El Toro and MCAS Tustin to Miramar, California, the redeployment of Marines after operations in Kosovo, and Marine presence in Istanbul, Turkey, providing humanitarian assistance and disaster relief in the aftermath of an earthquake.[224]

Marines conducted operations in East Timor in support of Australian-led International Forces to provide heavy-lift assets in response to the crisis associated with political unrest, and exercises in Greece and in Odessa, Ukraine, that simulated combined peacekeeping missions with soldiers from other NATO countries (e.g., France, United Kingdom, Turkey). Marines from Camp Pendleton, California, battled wildfires in Salmon-Challis National Forest, Idaho for four weeks.[225]

In 2001, General Jones published "Marine Corps Strategy 21," providing the Marine Corps with an overarching operational doctrine that would guide force structure, training and education, and acquisition.[226]

On September 11, 2001 at 9:38A.M., terrorists using commercial aircraft attacked, among other sites, the Pentagon in Washington D.C. Marines

223. See Appendix E.
224. DON, HQMC, *Marine Corps Command Chronology, 1999-2003.*
225. *Ibid.*
226. DON, HQMC, *Marine Corps Strategy* 21, 2001.

played a significant role in rescue and recovery efforts following the attack, by responding immediately with combat air patrols over the homeland. Marines also deployed to Pakistan as part of *Operation Enduring Freedom* in support of combat operations against Taliban and al-Qaida forces operating in southern Afghanistan.[227]

While continuing to conduct combat operations in Afghanistan, the Marine Corps conducted the largest urban-warfare exercise in its history as part of the larger exercise called Millennium Challenge 2002. This exercise included over 1,200 Marines and other forces that focused on warfare in urban terrain. As training exercises such as Millennium Challenge were ongoing, President George W. Bush formalized his warning to Iraqi leader Saddam Hussein, calling for the leader to eliminate all weapons of mass destruction or face military action from the United States, through a joint Congressional resolution authorizing the president to use military force.[228]

In the fall of 2000, Marines supported various operations on a global scale. The 2nd Marine Division led the task force Horn of Africa against terrorism, by disrupting terrorist cells that operated in the region. Marines of I Marine Expeditionary Force (I MEF), under the leadership of Lieutenant General James T. Conway, deployed to the Middle East and established the I MEF headquarters prior to the start of combat operations against Iraq.

In early 2003, large groups of Marines began to deploy to the Persian Gulf as the Department of Defense started to build combat power in the region for operations against the forces of Saddam Hussein. The president continued to send more troops to the Middle East, with the total of U.S. forces in the region reaching nearly 150,000 troops.[229] On January 13, 2003, General Jones turned over his command responsibilities as

227. *Command Chronology, 1999-2003.*
228. *Ibid.*
229. *Ibid.*

Commandant to General Michael W. Hagee, during a ceremony at the U.S. Naval Academy. In addition to serving as Commandant, General Jones would become the first Marine to lead the U.S. European Command and serve as the Supreme Allied Commander, Europe.

The Interview

General Jones was interviewed for 45 minutes at his place of business, Jones Group International, in Vienna, Virginia. When asked how the command climate influences the ethical behavior of the Marines, he replied: "Totally. I mean, if you don't have a good command climate you don't do anything well, let alone ethical behavior." The general emphasized three areas during the interview: setting the example, discipline, and understanding the environment.

This section begins by discussing the importance of setting the example and the importance of guidance offered by the commander. This is followed by a discussion on discipline and how the commander can influence ethical behavior and reinforce core values through martial arts training. Finally, the section concludes with a discussion on the significance of the environment relative to the visibility of commanders and its influence on ethical behavior.

Setting the Example

One of the most important steps a commander can take to influence the ethical behavior of Marines is to lead by example. As General Jones put it, "the first thing that you have to do is obviously lead by example, by how you handle yourself ethically." General Jones discussed the importance of troops seeing the commander behaving ethically and leading by example.

General Jones was influenced by the behavior of others, especially his father, a World War II Marine, and his uncle, who retired as a Lieutenant General from the Marine Corps in 1973. General Jones highlighted the importance of setting the example when he quoted his late father, who always reminded him of the maxim "officers eat last." In other words, it is the responsibility of officers to take care of the troops and always to put the needs of the troops before their own. While growing up, General Jones had good role models and examples to follow. These individuals lived the core values of honor, courage, and commitment, and stressed the ideas of selflessness, consistency in behavior, and upholding one's values.

He also discussed the importance of open communication and consensus building as part of the command climate that promotes ethical behavior. The commander must set the tone for the climate to influence the ethical behavior of the Marines. This must be established up front: "That kind of guidance comes from the top. You have got to make that part of what you expect your subordinates to deal with." For the climate to influence ethical behavior, the commander must create an environment where subordinates feel they can voice their opinions. This invitation from the commanding officer must be aimed at every level within the command. General Jones stressed that "you have to provide the command climate that generates those (ethical) discussions, not from the top down, but really from the bottom up." In other words, the NCOs and the junior officers should conduct ethical discussions on a regular basis, which reinforce the Marine Corps' core values.

If the core values are reinforced on a routine basis at the lowest possible level of command, the likelihood of reducing misconduct incidents increases. According to General Jones, "you must keep reminding people because new people are coming in all the time. If they don't hear that, then suddenly, the first thing that happens is you will have a major training or a major hazing incident." In addition to open communication, General Jones suggested that the commander should attempt to get buy-in from the subordinates relative to the decision making process within the command.

As part of the command climate and open communication, the general discussed the value of consensus building. As the commander becomes more senior (e.g., a colonel or general officer), consensus building takes on an even greater level of importance in establishing the right command climate. The normal tenure for any commander in the Marine Corps is two years. Therefore, to have a climate that is enduring, the commander must build a consensus among his staff, subordinate commanders, and senior enlisted leaders to ensure they "agree that what you're asking them to do is worth doing." When commanders build a consensus, they receive buy-in from the subordinate leaders tasked to carry out the commander's guidance. This makes people feel that they are part of the decision making process and they have a greater feeling of ownership in the product. Furthermore, consensus building might tend to mitigate poor decisions made in a vacuum. Examples include instances in which people take shortcuts to improve efficiency due to a lack of good communication with their seniors, or when someone is hesitant to report an ethical issue that senior leadership needs to address.

General Jones gave the example of a person who received their guidance and six months later was able to implement that guidance as if it were their own idea. He argued that "successful leadership is when the feedback you get indirectly or even accidentally confirms what you thought about six months before or a year before is actually working. So, it is the same thing with ethical behavior. And this is something you must talk about."

Discipline

General Jones listed some indicators that reflect a climate focused on ethical behavior. One of the key indicators is discipline: "What do the troops look like? How do the officers handle themselves? What is their off-duty incident rate? How are they doing on drugs and alcohol?" General Jones suggested that discipline is critical to operational confidence and success on the battlefield. He brought up the Marine Corps Martial Arts Program (MCMAP) as one of the primary tools for reducing misconduct incidents and poor ethical decision making within the organization.

It is important for commanders to build an ethical foundation for the Marines. General Jones stated that "understanding ethical behavior within the rules of engagement for example, on the battlefield, is extremely important." This ethical foundation is based on the Marine Corps' values and discipline. One of the programs General Jones implemented when he was Commandant to instill discipline and reinforce Marine Corps values was the MCMAP.

When General Jones was a second lieutenant serving in Vietnam, he was assigned to work with the Korean Marines, who would get up in the morning and exercise using the martial art of Taekwondo. This type of martial art required discipline and was physically demanding. "I saw the discipline and ethical value of it," recalled General Jones. "So, in 1996 at Camp Pendleton, I made it voluntary in the battalion. Almost immediately, the ethical behavior of the battalion changed."

General Jones realized that Marines would embrace the idea of being confident enough to defend themselves, avoid confrontation, and, if necessary, fight without losing their dignity. As a battalion commander, he noticed a reduction in drug and alcohol-related incidents and reckless behavior. He also realized more reenlistments and evidence of discipline in the command. Based on his experiences, when he became Commandant he directed the Marine Corps to adopt the MCMAP Corps-wide.

Today, Marines practice MCMAP in all Marine Corps units. A MCMAP belt is highly valued by Marines. The MCMAP belt is a "badge of courage" that not only signifies a certain level of martial arts proficiency (e.g., tan belt for beginners, and green belt for advanced), but also demonstrates a high degree of self-discipline. "They don't want to lose that belt," General Jones said, "And they could lose that belt just by being undisciplined and, in many cases, unethical." General Jones suggested that when commanders institute a strong MCMAP within their units, they will realize a reduction in misconduct incidents associated with poor ethical decision making, and they will benefit from an increase in the overall discipline within the organization.

The Environment

General Jones argued that "it's much tougher now to command. It's much tougher to be a CEO of a corporation. Anybody who thinks they can hide behind their behavior is wrong." Due to social media and enhanced technology, commanders are much more visible today than in years past. Consider the CNN effect: "So, the first thing I think you have to do in training our officers and staff noncommissioned officers is to make sure that they understand that they are not invisible. In fact, they are visible 24/7 and 365."

Since commanders are always being watched, the example they set is critical. Due to the advent of social media, commanders are very visible, and they must be very careful about what they say, what they post, and what personal information they decide to share because they represent the Marine Corps as an institution. General Jones believes that an increased level of visibility is a good thing to hold commanders accountable for their own leadership by example. Every move a commander makes is scrutinized: "You should never do anything that embarrasses yourself, your family, your unit, or your country." Setting the right command climate that influences the ethical behavior of the Marines takes effort, and constant attention: "You have to work on it." Again, he used the example of hazing.

When General Jones was Commandant, he determined there would be, "only one form of authorized hazing, and that is recruit training and officer training. That's it." During the interview, he mentioned that the problem of hazing and other issues is like ethical training, which must be discussed on a regular basis or the Marines' character development will begin to erode. He also suggested that hazing is something that occurs when leaders have lost control of their organizations based on a lack of attention to getting out and seeing what is going on.

Summary

General Jones believes that one of the most important things a commander can do in establishing the right climate is to set the proper example. He suggested that commanders must adopt open communication as part of their command climates to promote ethical behavior. He recommended consensus building for commanders to ensure that they have buy-in from other leaders within the organization. General Jones encourages commanders to study various metrics such as alcohol-related incidents, drug use, and other misconduct as indicators of the command climate's success.

According to General Jones, one of the best tools a commander can use to influence the ethical behavior of the Marines is the MCMAP. He discussed how the program promotes ethical behavior through the discipline associated with martial arts. He pointed out that Marines cherish the MCMAP belts that symbolize an individual's level of martial arts proficiency and discipline.

Finally, General Jones discussed the environment, and how it acts as a double-edged sword given that a commander's behavior is always being scrutinized and social media opens the door for an even greater amount of inspection. Conversely, social media can also ensure that commanders are being held to their own high standards while at work, and when they are living their private lives away from the base.

General James Jones and the author.

GENERAL MICHAEL W. HAGEE, 33RD COMMANDANT (2003-2007)

Background

On January13, 2003, General Michael W. Hagee became the 33rd Commandant of the Marine Corps during a ceremony at the U.S. Naval Academy.[230] At the time there were approximately 177,000 Marines serving on active duty in the Marine Corps. While General Hagee assumed command, Marines were rapidly building up combat power in Kuwait, as the United States faced a possible war with Iraq. In March 2003, *Operation Iraqi Freedom* began with cruise missile attacks on military targets around Baghdad. In early April, Iraqi resistance in Baghdad collapsed as American forces captured the capital city, ending major combat operations.[231]

Other significant events took place after hostilities in Iraq ended. In the spring of 2003, Marines left Vieques, Puerto Rico and Iceland after years of a Marine Corps presence in both locations. Marines were notified that they would be returning to Iraq as part of the U.S. troop rotation to fight a growing insurgency in the country. While some Marines were preparing to return to Iraq, others were pursuing al-Qaida and Taliban forces in Afghanistan.[232]

In 2004, the Corps participated in multiple operations, including evacuating American citizens from Haiti as part of a multinational interim force, conducting security and humanitarian assistance operations with the new Iraqi security forces, providing humanitarian assistance and disaster relief in the Philippines after two weeks of severe tropical storms, and participating in humanitarian assistance operations for victims of the tsunami that swept across the Indian Ocean in December 2004.[233] Another significant event occurred in 2005—the creation of a

230. See Appendix E.
231. *Command Chronology, 2003-2007.*
232. *Ibid.*
233. *Ibid.*

new component within Marine Corps' Special Operations Command (SOCOM), known as MARSOC. This organization would provide SOCOM with a Marine Corps component.

In 2006, Marines continued to support operations in both Iraq and Afghanistan.[234] In the spring of 2006, the Marine Corps began moving personnel and equipment from Okinawa, Japan to Guam, as part of the Pentagon's larger effort to better align security forces in the Pacific Rim. Other significant events in 2006 included the evacuation of U.S. citizens from Beirut, Lebanon, due to fighting between Hezbollah and Israel. The Wounded Warrior Center opened at Camp Pendleton to mirror the center at Camp Lejeune. Finally, the National Museum of the Marine Corps in Quantico, Virginia was officially dedicated on the Corps' 231st birthday. In January 2007, General Hagee retired from the Marine Corps after serving for 42 years.[235]

The Interview

General Hagee was interviewed for approximately 90 minutes at his place of business, the Admiral Nimitz Foundation, in Fredericksburg, Texas. Upon entering the general's office, he immediately asked me to come over to his desk to examine the picture of a small boy standing next to an old man. General Hagee asked me if I knew who the boy was in the picture and I said I did not know who it was. He proceeded to tell me that the boy was Chester Nimitz and the old man was his grandfather, the man who raised him. This led to a discussion about Admiral Nimitz and how he was the epitome of character and a person who set the right command climate for his organization.

General Hagee emphasized the importance of setting the example, good communication, holding people accountable, and educating Marines on core values. This section addresses setting the example and how

234. *Ibid.*
235. *Ibid.*

command climate is not a function of the commander and his or her success. Rather, the command climate is determined by the people in the organization and preserving the integrity of the institution. A discussion on open communication and how it builds trust that promotes ethical behavior within the organization follows. The section then moves to the importance of maintaining high ethical standards and holding people accountable. It concludes by addressing General Hagee's approach to educating the Marines on core values using the Socratic method and using ethical dilemmas to influence the ethical behavior of the Marines in the unit.

Setting the Example

When General Hagee was asked what had the greatest influence on him and his approach to command climate, he replied, "Watching individuals, both good leaders, guys that set the right climate, and individuals that did not set the right climate. So, it's constantly looking at your environment and learning from both good and bad examples." He referenced Admiral Chester Nimitz and how this renowned officer always left his ego at the door—an important aspect of setting the right example.

General Hagee described the command climate set by Nimitz as the textbook example of an officer who was more concerned about the institution and other people than with his own personal success. He shared two stories that can be used by commanders today of how Nimitz set the example and developed the right command climate for the Pacific Fleet during World War II.

In 1941, Nimitz led the Bureau of Navigation in Washington, D.C. His position required close contact with the Chief of Naval Operations and the White House. President Franklin D. Roosevelt had become frustrated with the Commander in Chief, U.S. Fleet, Admiral John Richardson, whom he relieved in early 1941.[236] Because of his positive reputation in Washington, Roosevelt asked Nimitz to be the new fleet commander in place of Richardson. Nimitz turned down the promotion because he believed it would not be good for the Navy to pass over a hundred or more senior admirals who should be considered for the position.[237] Nimitz was more concerned about the Navy as an institution and the other more senior admirals than he was about himself. This attitude was typical of Nimitz, as evidenced by the command climate he established in the Pacific Theater. General Hagee recommended that I read *Nimitz*, a biography on the admiral.

General Hagee discussed the importance of the commander "checking his ego at the door" and focusing on the institution and the people within the organization. These are critical to influencing the ethical behavior of the Marines. He said "it's not about the individual (commander). The guys I told you about who got fired, it was about them. Admiral Nimitz? It was never about him." General Hagee mentioned one final example of Nimitz setting the right climate and leading by example.

After the attack on Pearl Harbor, Admiral Husband E. Kimmel was relieved of his command of the Pacific Fleet at Pearl Harbor, and President Roosevelt directed Admiral Nimitz to assume the duties as the Commander in Chief of the U.S. Pacific Fleet.[238] As one of his first tasks, Nimitz had to select his new staff in Hawaii. Kimmel's staff was expecting the worst and was preparing to be dismissed just before Christmas. Nimitz gathered Kimmel's entire staff together and told them he desired to keep them on for continuity and that if anyone requested a transfer, he would listen to them and try to help them to get the assignments they requested.[239] "That's really command climate," according to General Hagee. In response to the question, "How does the command climate influence ethical behavior?" He pointed to the importance of open communication and holding people accountable.

236. Potter, E.B. Nimitz. Annapolis, MD: Naval Institute Press, 1976.
237. *Ibid.*
238. *Ibid.*
239. *Ibid.*

Open Communication

One of the first things General Hagee mentioned about communication is that commanders must have the ability to communicate with their Marines on all levels. He used an example from when he was a platoon commander in Vietnam. A white Marine from Alabama hung a small Confederate flag on his tent, and a black Marine took offense to this gesture. General Hagee allowed the two Marines to discuss the issue in a civil manner, and once the white Marine understood the black Marine's position, he removed the flag from the tent. In other words, General Hagee concluded: "The Marines provided the solution for the betterment of that command, and therefore, the command climate. Maybe we can bring problems to him (the commander), and he is going to help us resolve those problems." General Hagee suggested that this type of communication promotes trust and facilitates better ethical decision making by the Marines.

Regarding a commanding officer's, staff, General Hagee believes that "you have to get the staff to where they will open up with you and have a conversation with you. And that means you must keep your damn mouth shut and listen, and not shoot anyone in the face." During the interview, he gave as an example the ethical dilemmas many senior commanding officers face regarding the perquisites, benefits, and other advantages that are sometimes made available to senior leaders.

When he was the commanding general of First Marine Expeditionary Force (I MEF), General Hagee had to take a flight to Twenty-nine Palms, CA. There was an empty seat on the plane, and he could have easily used that seat to bring his spouse on the trip. His staff recommended that he not use the extra seat for his spouse, as it would set the wrong example and violate ethical rules associated with the conduct of senior leaders. The general, of course, listened to his staff and did not take his wife on the trip. This is an example of listening to the staff and exercising good leadership by example. Another instance of open communication that

Hagee shared in the interview took place during the planning phase for the invasion of Iraq.

During I MEF's planning leading up to the invasion of Iraq, General Hagee's staff developed a plan for the ground invasion. Because of the type of climate General Hagee had established within the I MEF headquarters, Marines were able to feel comfortable voicing their opinions and disagreements with various planning suggestions made by senior leaders. On one occasion, General Hagee was being briefed on the final plan, which had been in the works for three to four months, and a captain (junior officer) told General Hagee that he did not think the plan would work. General Hagee replied, "ok, now tell me why?" The captain explained his position in a very articulate manner and General Hagee agreed to change the plan based on the captain's recommendation. This is an example of a command climate that is based on trust, open communication, and people trying to do what is in the best interests of the organization.

Accountability

General Hagee stressed the importance of holding people accountable. One example he discussed occurred during his time as the commanding officer of the First Marine Division at Camp Pendleton. A Marine was shooting horses in California, and General Hagee was going to court-martial the Marine. General Hagee received a phone call from a senior leader in the office of the Secretary of Defense who suggested that the general discharge the Marine without a court-martial. This was an instance of political pressure from the parents of the Marine, who were vying to get the Marine's sentence reduced. General Hagee was informed by his legal officer after his decision to court-martial the Marine was finalized, that the entire division was waiting to see if he was going to hold the Marine accountable for his actions. "That," remarked General Hagee, "becomes very important in setting the climate." He suggested that junior Marines admire their commanders and want to be like them, which is a positive aspect of command climate. However, the negative aspect of climate is when Marines recognize that the commander is serious and will hold them accountable if they make mistakes of character: "If I cross that ethical line, then he (the commander) is going to hold me accountable, and I don't want to do that."

In other words, a commander cannot expect Marines to make ethical decisions if they, as the commander, do not uphold the Corps' high standards and hold those accountable who choose to violate them. One way for the commander to mitigate the risk of Marines not following the standards is to consistently teach and reinforce the Marine Corps' core values.

Education and Values

General Hagee's approach to climate command was influenced by his tour as an instructor at the Naval Academy when he was a colonel. The Navy experienced a cheating scandal, and the Secretary of the Navy sent Colonel Hagee to the Naval Academy to "fix" the ethical problem. He suggested to the staff that the best way to tackle the problem was to establish a character development program for all Midshipmen. The survey professors at the Academy took umbrage with the plan until he asked them. "What happens if you shave a little bit to save money on that bridge you're making? Is that an ethics thing?" The professors agreed that ethics must become part of the instruction at the Academy. The Academy subsequently established a four-year ethics program across the curriculum, which is still taught today.

General Hagee discussed how core values must be reinforced on a regular basis by the leaders at the lowest possible level, to have the greatest influence on the Marines in the unit. He stressed the importance of discussions on ethical dilemmas: "That is part of helping to set, in my opinion, the command climate to where individuals can think through that—and by having that discussion, you set up better individuals that can make better decisions, which is going to help the command climate."

General Hagee disagrees with the approach in which an officer lectures Marines about classical ethical theory (e.g., Immanuel Kant, Utilitarianism, or Aristotle): "You've got to either have a real problem that you're trying to resolve, or you have an issue that you want them to talk about and say, if that issue came here, how would you handle it." According to General Hagee, one of the best ways to approach discussions on ethical dilemmas is to use the Socratic method.

The Socratic method is "a pedagogical technique in which a teacher does not give information directly but instead asks a series of questions, with the result that the student comes either to the desired knowledge by answering the question or to a deeper awareness of the limits of knowledge."[240] The Socratic method enables the leader to guide the discussion without giving the answer to the others. It forces people to think through the problem, especially during periods of awkward silence, when others are not sure how to respond to an ethical dilemma. As General Hagee pointed out, "Being yes-no, black-white, all the time, in my opinion, does not set the type of command climate that you necessarily want. Because there are ethical dilemmas." He suggested that the Marine Corps should teach the ethical dilemma associated with the story of Oliver P. Smith and the "Chowder Society." He recommended that I read a biography of Smith, *For County and Corps, The Life of General Oliver P. Smith.*

After World War II, Congress was seeking for the Marine Corps to be subsumed into the Army. A small group of Marine Corps officers along with a few others who called themselves the "Chowder Society" joined together to write the "bended knee" speech for then Marine

240. *The American Heritage Dictionary*, 2018.

Corps Commandant, General Vandegrift. The speech aimed to explain to Congress why the Corps should remain "a force in readiness, with its supporting arms and air wings intact."[241] Based on this effort, the resulting National Security Act of 1947 wrote into law the mission of the Marine Corps.[242]

O.P. Smith, a career combat leader during World War II and the Korean War, chose not to be a part of this secret committee, as he believed they were not taking an ethical approach to securing the future of the Marine Corps.[243] Based on his lack of participation in this secret group, "Smith became an outsider and an enemy, by not agreeing with the group's methods, even though he fully supported their objective."[244] As General Hagee put it, "O.P. Smith refused to participate because he thought it was underhanded." Which side was correct, O.P. Smith's view or the view of the Chowder Society? General Hagee suggested that commanders bring these types of discussions to their Marines on a regular basis, to reinforce the core values by using the Socratic method, forcing Marines to think through difficult ethical problems that do not always have easy solutions.

Summary

General Hagee discussed the importance of setting the right example when you are leading others. He mentioned that the leader's focus should be on the people and the institution, not on the leader. He referenced Admiral Chester Nimitz as the epitome of selflessness and as a leader who was focused on the people and the Navy as an institution.

General Hagee emphasized the need for open communication if commanders hope to create a command climate that influences the

241. Shisler, G. *For Country and Corps, The Life of General Oliver P. Smith*. Annapolis, MD: Naval Institute Press, 2009, p. 117.
242. *Ibid.*
243. *Ibid.*
244. *Ibid*, p. 120.

ethical behavior of Marines. Open communication must be part of the climate of the organization to get Marines to talk to each other in a civil manner and to facilitate recommendations from subordinates without fear of repercussions.

General Hagee discussed the value of high standards and holding Marines accountable if they violate those standards. He offered the Socratic method as the best way for commanders to ensure that Marines are having discussions about ethical dilemmas associated with the Corps' core values. These discussions, General Hagee recommended, must be led by junior officers and NCOs at the lowest possible level of command, to really influence the ethical behavior within the organization. He alluded to a historical figure, General O.P. Smith, as a good example for commanders to use when discussing ethics with their Marines.

General Michael Hagee and the author.

GENERAL JAMES T. CONWAY, 34TH COMMANDANT (2006-2010)

Background

General James "Jim" Conway became the 34th Commandant of the Marine Corps on November 13, 2006, during a ceremony at Marine Barracks, Washington, D.C.[245] At the time, there were approximately 179,000 Marines serving on active duty in the Corps. General Conway's number one priority was to support the Marines and Sailors in combat. In late 2006, Marines were experiencing a decline in fighting in Iraq as a result of "The Awakening" (Sunni Sheiks cooperating with the Iraqi Police and Iraqi Provisional Army to counterbalance Al-Qaeda). At the same time, the Taliban in Afghanistan were becoming a significant problem requiring more Marines to be deployed to that country.[246] Due to the number of casualties returning home, General Conway ordered the establishment of the Wounded Warriors Regiment at Marine Corps Base Quantico, Virginia. This organization would assist injured Marines through their recovery and transition to their next assignment or to the civilian life. The two subordinate Wounded Warrior Battalions were previously established on each coast of the United States.[247]

In 2007, Marines continued to participate as part of the NATO-International Security Assistance Force. Simultaneously, Marines increased their footprint in western Al Anbar Province in Iraq. During this period, General Conway visited Marines in Iraq and Afghanistan and conducted town hall-style meetings to discuss numerous topics of interest. This style of leadership was very indicative of the Commandant's command climate.[248]

245. See Appendix E.
246. DON, HQMC, *United States Marine Corps Concepts and Programs*, 2009.
247. *Command Chronology, 2006-2010.*
248. *Ibid.*

In September of 2008, the Marine Corps continued to support combat operations in both Iraq and Afghanistan, with the Iraqis taking over much of the security force mission in their country. In Afghanistan, Marines captured key areas from the Taliban. During this period, Barack Obama was elected as the 44th President of the United States of America.[249]

In 2009, Marines continued to support both *Operations Enduring/ Iraqi Freedom* in Afghanistan and Iraq. Marines participated in numerous operations that included combat operations, stability, and security operations, and transitioned to counterinsurgency operations. Marines transferred responsibility to Iraqi forces in Anbar Province, while Marines in Afghanistan continued to fight the Taliban in support of the Afghan National Army.[250]

During this period, the Corps was deployed globally, providing support to many other nations and allies through theater security cooperation efforts. In addition, Marines were deployed in support of civil-military and humanitarian assistance operations throughout the world, as in the Philippines with disaster relief in response to the devastating effects of two tropical storms.[251] Other examples of disaster relief and humanitarian operations included disaster relief in Haiti, where Marines responded to the massive earthquake, and in Pakistan.[252] Coinciding with General Conway's change of command, Marines were once again deployed to the Philippines to help in the response to Typhoon Megi, in October of 2010. They provided food, water, tents, supplies, and air support to isolated villages that were destroyed by the typhoon. On October 22, 2010, General Conway relinquished his post as Commandant during a ceremony at Marine Barracks, Washington, D.C. to General James F. Amos.[253]

249. *Ibid.*
250. *Ibid.*
251. *United States Marine Corps Concepts and Programs*, 2009.
252. *Ibid.*
253. *Command Chronology, 2006-2010.*

The Interview

General Conway was interviewed at his residence for approximately 90 minutes. General Conway emphasized three areas during the interview. First, that the climate is set by the commander and expressed through his or her philosophy of command. Second, that the commander must lead by personal example. And third, that commanding officers must act like human beings.

This section discusses General Conway's thoughts on the philosophy of command, particularly the importance of conveying the commander's expectations to Marines through their commitment to the core values. The section also addresses General Conway's thoughts on the need for open communication, the concept of collective leadership, and the importance of receiving constant feedback. This is followed by a discussion of leadership by example. During the interview, General Conway pointed to the need for the commander to instill discipline by holding people accountable and taking care of them while looking out for the institution. Finally, the importance of acting like a human being by taking care of the Marines' families and by having a sense of humor is reviewed.

Philosophy of Command

General Conway discussed the philosophy of command as one of the first things a commander needs to convey to Marines. This philosophy must be delivered to all the Marines (e.g., officers, NCOs, the commander's staff, and the troops) face-to-face. "You have to talk to people," General Conway affirmed, "You have to be able to read the body language. The whole thing of communication up front, I think, is really key." The commander must practice what he or she preaches: "You can't say one thing and then do something else." The commanding officer must walk the talk and follow through with those areas covered in the philosophy of command.

An example of General Conway's philosophy of command was his 34th Commandant's Planning Guidance,[254] where he identified seven areas of focus for the Marine Corps, which included an emphasis on the core values and creating quality citizens. In 2007, he published a communication plan in which he stated that "Assuring Marine Corps success requires effectively articulating what we stand for, what we do, who we are, and what we aspire to be."[255] As part of his planning guidance and his communication plan, the Commandant emphasized the importance of character and creating quality citizens, which stem from the Marine Corps' core values.

General Conway put forth a document that established his expectations for the Marine Corps, especially his expectation that the Corps would endeavor to rededicate to the core values.[256] As Commandant, he set the tone on the importance of values and the responsibility of the commanding officer to establish clear standards of personal behavior for their organizations, as well as maintaining those standards.

Another example of General Conway's commitment to core values and the importance of an ethical command climate was his direction to all commanders on the topic of sexual assault: "I charge you with educating yourselves on sexual assault, identifying the misconceptions your Marines may have about it and creating a command climate that dispels these myths and establishes clear standards for personal behavior."[257] In other words, it is the responsibility of the commander to set the right climate that promotes ethical behavior in accordance with Marine Corps

254. Conway, J. "34th Commandant of the Marine Corps: Commandant's Planning Guidance." *Leatherneck Magazine*, 90 no. 1 (2007b): 12-25.
255. DON, HQMC, *Strategic Communication Plan*, 2007.
256. Conway, J. "34th Commandant of the Marine Corps: Commandant's Planning Guidance." *Leatherneck Magazine*, 90 no. 1 (2007b): 12-25; DON, HQMC, Marine Corps Vision and Strategy 2025.
257. Conway, J. "Commanders' Responsibilities in Addressing Sexual Assault." White Letter No. 02-09. Headquarters Marine Corps, Washington, D.C., 200, p. 1.

values. General Conway made it clear that "ethical behavior is a function of leadership."[258]

As part of the command philosophy and the climate set by the commander, commanders must have open communication within their units to influence ethical behavior. The relationship between the commander and the subordinates must be built on trust and frankness, which are facilitated by open communication. This promotes collective leadership, resulting in good feedback to the commander. General Conway shared one example that incorporates open communication, trust, and feedback from the time he was in command of I MEF in Iraq.

Brigadier General Richard Natonski arrived from Camp Lejeune, N.C. in command of Task Force Tarawa and was assigned to be attached to I MEF. Natonski claimed that his task force could do anything that General Conway asked him to do. He was so eager to accept tactical missions that General Conway began to wonder if Natonski was a "yes man," or just truly enthusiastic for work. To satisfy his curiosity, General Conway sent Natonski a ridiculous assignment, then called him and asked "what do you think?" There was a pause on the other end of the phone, and Natonski said, "Sir, I think that's the most fucked up thing I ever heard in my life." General Conway told Natonski that he had just passed the test!

In other words, commanders must set a climate that promotes honest and open communication, where people can speak their mind without fear of repercussions from the senior leadership. This type of climate, where the commander is very approachable, promotes ethical behavior in organizations. General Conway actively sought the ideas of others prior to making decisions. He valued the concept of collective leadership.

"I always believed that other people's thoughts count," General Conway said, "Early on I realized the value of democratic or collective leadership."

258. Conway, J. "Contacts with Representatives from Industry." White Letter No. 08-07. Headquarters Marine Corps, Washington, D.C., 2007, p. 2.

General Conway believes that a commander must establish a climate that promotes buy-in and develops trust among subordinates. Open communication coupled with collective leadership promotes trust and confidence, and results in a greater likelihood of ethical behavior from subordinates. As General Conway said: "You really do want to hear what they have to say, and you really do want to listen. It is important."

When General Conway was Commandant, he had concerns about how to take care of wounded warriors. A member of his staff recommended the establishment of a Wounded Warrior Regiment that could oversee the two Wounded Warrior Battalions. Following this advice, General Conway set up the Wounded Warrior Regiment in Quantico, Virginia. He always looked for the good ideas of others: "One of the tenets of a good leader is to be able to understand it, appreciate it, and then put it into effect. So, that's command climate."

As part of open communication, and closely associated with collective leadership, General Conway emphasized that commanders need to receive continual feedback to ensure their command climate is working. As he put it, "body language with your commanders, body language with your staff, willingness of people to give feedback" are all indicators of climate effectiveness and its influence on the ethical behavior of the Marines.

If the commander asks Marines what they think, and the climate is of a nature where people feel intimidated, then the conversations will be very short and will rob the commander of the opportunity of hearing good ideas from subordinates. General Conway suggested that commanders can receive valuable informal feedback by walking around: "You can tell when troops are happy and when they're not. You can tell in your individual conversations with them if they feel they can talk to you." He mentioned that when commanders ask their Marines how they are doing and solicit their opinions and recommendations, Marines will be more inclined to make better ethical decisions, because of the trust and confidence that has been established within the organization between the leader and the subordinates.

Another example of taking care of Marines was how General Conway handled his general officers' professional development. General Conway had been a general officer for nearly ten years before he became Commandant. He was counseled on his performance as a general only very rarely. Therefore, as Commandant, he initiated a process where all one and two-star general officers would be counseled by the general officer senior to them in their chain of command. This report would be sent via the three-star level to the Commandant notifying him that all generals were counseled. This process promoted open communication, trust, and good leadership. Senior leaders were willing to look other senior leaders in the eyes and discuss with them their performance and their opportunities for future advancement.

Leadership by Example

General Conway discussed two areas integral to leadership by example related to command climate and its influence on the ethical behavior of the Marines. First, commanders must instill discipline in their Marines by holding them accountable for their actions. Second, it is the responsibility of the commander to take care of the Marine while protecting the Marine Corps as an institution.

"Discipline is important in a unit," General Conway said, "You want to create a good command climate where people know that you're going to get rewarded for good behavior, but you're going to be punished for bad behavior." He discussed the importance of setting high standards and holding people accountable to those high standards. He suggested that it is the responsibility of the commander to discipline Marines when necessary and that a commander cannot be worried about whether he or she is popular with the troops: "with [command] comes the responsibility of disciplining when you have to. You can't be reluctant to do that."

One example of this is when General Conway was a company commander and he had to discipline two of his best NCOs. General Conway's Marines were training with German troops in the Black Forest. When the training concluded, two NCOs met a German boy who put them in touch with a merchant who sold them peach brandy. Prior to the hike back to the ships, the two NCOs were drunk and were barely able to walk back to the ships with the rest of the company. General Conway held the NCOs accountable by fining them and putting them on restriction, which precluded the Marines from seeing England during the next port-of-call. General Conway recalled that the word spread like wild fire. "If the company commander is willing to take two of our best guys and hammer them, gosh, God forbid if we do something stupid on liberty."

General Conway was the only commander who had no incidents while on liberty in South Hampton or in Liverpool. As he put it, "people take on the personality of the commander and reflect it." He highlighted the importance of leadership by example and the need for holding people accountable. One final example of accountability is when General Conway had to relieve two of his battalion commanders due to abusive leadership styles.

He became aware of the problem through junior officers, who expressed that the battalion commanders were out of control and abusive. General Conway also uncovered more information about the commanders through division inspections and command climate surveys, which provided him with enough information to hold the commanders accountable, relieving them of their duties.

General Conway believes it is a commander's responsibility to protect the institution. To demonstrate this point, he used two examples relevant to climate and ethical behavior. The first example dealt with one of General Conway's general officers. The second example juxtaposed the Commandant's concerns for a warfighting institution, against political agendas being set in Washington.

When General Conway was Commandant, he had to relieve one of his general officers due to misconduct. General Conway referred to it as "the Lance Corporal Rule": "If a lance corporal gets punished for doing this, then a general gets punished for doing the same thing. It's just that simple." General Conway had to run the disciplinary action through the Secretary of the Navy, who was opposed to the punishment for political reasons. General Conway approached the Secretary on three different occasions to attempt to convince him to do the right thing. Eventually, he was successful in convincing the Secretary to agree with the dismissal. The general was relieved and forced to retire as a colonel. General Conway pushed to do the right thing for pure ethical reasons, but also to do the right thing for the institution, based on the negative precedence that would be set if the officer kept his position or retired as a general. The second example dealt with the U.S. Congress.

When General Conway was in Washington, D.C., numerous political initiatives would appear, such as homosexuals serving openly in the military or women joining infantry units. These initiatives had an impact on the Marine Corps as a warfighting organization. General Conway was opposed to these initiatives: "I think it's really hard to put a Marine into a barracks with a homosexual, or two, or three, and think that, okay, there's going to be automatic bonding here."

When he looked at each of the proposals, he asked, "does that make the Marine Corps a better fighting force?" If the answer was yes, then it was something the Marine Corps should do. But if the answer was no, then it was not in the best interest of the Marine Corps as an institution. As he put it, "As far as society and equal, absolutely. But that's different; that's not fighting and dying. That's just living." In other words, General Conway never looked at things from the perspective of what's best for Jim Conway, or society. Rather, he considered ideas and tried to determine what was best for the United States Marine Corps as a warfighting institution.

Being Human

General Conway discussed the significance of acting like a human being when leading others: "I don't think I ever held a school circle—that's when you gather all the Marines in one place and break 'em down, and again you go eyeball to eyeball with people—that I didn't feel like I was trying to reveal some of my personality, some of my humanity." He addressed two areas relative to climate and its influence on the ethical behavior of the Marines. First, he mentioned the importance of taking care of a Marine's family. Second, he suggested that a commanding officer must have a good sense of humor.

Earlier, the Wounded Warrior Regiment was discussed as one of the examples of how General Conway cared for his wounded Marines. He recalled that his wife had been instrumental in that policy. "Hey, it's not just about the Marines," she reminded him, "It's about the families, too." General Conway always highlighted the importance of taking care of the families to his commanders. "While we recruit Marines," he said, "we retain families."[259] He believes that family readiness is instrumental to the long-term health of the Corps and to unit readiness prior to the rigors of deployments.[260,261] Commanders are responsible for ensuring

259. Conway, 2010 Conway, J. "Report to Congress on the Posture of the United States Marine Corps." Headquarters Marine Corps, Washington, D.C., 2010, page 9.
260. Conway, J. "34th Commandant of the Marine Corps: Commandant's Planning Guidance." *Leatherneck Magazine*, 90 no. 1 (2007b): 12-25."
261. Conway, 2010 Conway, J. "Report to Congress on the Posture of the United States Marine Corps. Headquarters Marine Corps." Washington, D.C., 2010.

their Marines' families are taken care of. This should be part of their command climate, which will also promote better ethical decision making because the Marines know the commander is looking out for them and their families.

At the end of the interview, General Conway advocated for commanders to have a good sense of humor. For example, when he was a company commander as a young captain, he told his company staff to schedule a range shoot. After the training was over and everything was cleaned up and accounted for, the company went into an administrative bivouac with hot chow and a campfire. "I remember walking around just standing back in the shadow," General Conway said, "but Jesus Christ, the humor and the levity that was going on with the harassment and the things that were being said. It was unbelievable."

This gesture raised morale and allowed the Marines to have fun while training in the field: "So, I think that that's an important part of creating a command climate and maintaining it, really that you can enjoy the humor that your troops put forward and be the butt of it if need be, but just have fun while you're doing it." General Conway summed up his thoughts on command, climate, and leading Marines with this notion: "There's something wrong with your approach if you can't enjoy the people around you that are patriots, dedicated, mission-oriented, and love their country."

Summary

General Conway recommended that commanders deliver their command philosophy to their Marines face-to-face, to ensure that all personnel understand the commander's expectations and intent. He believes the expectations for Marines should include a commitment to the Marine Corps' core values. In addition, he suggested that commanders create a climate that facilitates open communication where the commander is accessible, and one where every Marine can offer suggestions and recommendations to the commander.

General Conway suggested that commanders solicit feedback from their Marines and that commanders should seek the ideas and thoughts of others within the organization before making decisions. He believes in leadership by example. Commanders must establish a climate that instills discipline in Marines and holds them accountable for their actions. While commanders must take care of their people, they must also protect the Marine Corps as an institution, and they need to take care of their Marines' families. This will promote better ethical decision making, because the Marines know their commander is looking out for them and their loved ones. General Conway also advocated for commanders to have a good sense of humor and show their Marines that they are human beings who truly enjoy being a Marine and leading other Marines.

General James Conway and the author.

GENERAL JAMES F. AMOS, 35TH COMMANDANT (2010-2014)

Background

General James F. Amos became the 35th Commandant of the Marine Corps on October 22, 2010, during a ceremony at Marine Barracks, Washington D.C.[262] General Amos was the first Marine Corps aviator to become Commandant. He also became the first Assistant Commandant in 27 years who moved up to assume the duties as Commandant. At that time, there were approximately 202,000 Marines serving on active duty.

In 2010, General Amos issued his *35th Commandant's Planning Guidance*, with the number one priority of winning the fight in Afghanistan.[263] Commensurate with this priority, the Marine Corps conducted surge operations in Afghanistan under the command of Major General Mills and the 1st Marine Division. In keeping with his promise to keep the Marine Corps as the nation's Expeditionary Crisis Response Force, he deployed a Special Purpose Marine Air-Ground Task Force Crisis Response to both Africa and the Middle East.[264]

While Marines continued to fight the War on Terror, they also participated in numerous disaster relief/humanitarian assistance missions. These included the 2012 disaster relief operation in the Philippines, where Marines delivered relief supplies, and the 2013 operation during which Marines delivered supplies and humanitarian assistance in the aftermath of Typhoon Haiyan. Marines focused on assisting allied partners and friends through theater security cooperation efforts, by strengthening the military capacity of other partner nations such as Saudi Arabia, Kuwait, United Arab Emirates, India, Thailand, Australia, Japan, and the Philippines.[265]

262. See Appendix E.
263. DON, HQMC *35th Commandant's Planning Guidance*, 2010.
264. *Ibid.*
265. Amos, "Commandant's Report to Congress on Posture of the Marine Corps," 2014.

In 2013, Marines reduced their overall presence in Afghanistan due to the success in working with the Afghan National Police and Afghan National Army. As the Marine Corps decreased its presence in Afghanistan, it continued to contribute to global security and stability through Marine Embassy Security detachments and Fleet Anti-Terrorism Support Teams in support of U.S. diplomatic missions in Libya and Yemen, as well as the other 148 embassies and consulates throughout the world.[266]

Marines also participated in operations assisting American citizens at home, such as disaster relief in response to Hurricane Sandy. In addition to combat operations overseas and humanitarian assistance/disaster relief at home and abroad, Marines also conducted counterterrorism, foreign internal defense, and other special operations throughout the globe, including, combating al-Qaida organizations in Mali and Mauritania.[267]

The Marine Corps continued to provide expeditionary readiness in the Pacific through its forward presence in Australia. The Marine Rotational Force Darwin conducted training and readiness exercises throughout the Pacific region, focused on building partner capacity and theater security cooperation efforts with nations such as Thailand, Indonesia, and Malaysia.[268] In October of 2014, General Amos relinquished his duties to General Joseph Dunford, at Marine Corps Barracks, Washington D.C.

The Interview

General Amos was interviewed in March for 90 minutes. He was the only Commandant interviewed over the phone, due to scheduling challenges. During the interview, he expressed his view that the subject of command climate and its influence on the ethical behavior of Marines is critical, one of the most important things a commander does. As Commandant, General Amos linked good leadership with the spiritual health of the

266. *Ibid.*
267. *Ibid.*
268. *Ibid.*

Marine Corps. He approached the subject of command climate through a series of guidelines, papers, and face-to-face discussions (campaigns). These included: his *35th Commandant's Planning Guidance*, which discussed his priorities for the Marine Corps; White Letters on Command Climate and Leadership directed to commanding officers; the Heritage Brief, which provided details on what it meant to be a U.S. Marine, the importance of meeting the Marines Corps' high standards, and the expectations of the American people; and the Reawakening Campaign, directed at NCO leadership.

This section addresses command climate and leadership in terms of open communication, dignity, respect, accountability, and discipline. Subsequently, the significance of the Marine Corps' core values and the importance of the spiritual health of the Marine Corps are discussed. The key documents mentioned in the previous paragraph are used as references throughout the section.

Climate and Leadership

According to General Amos, "There is no more visible aspect of leadership than the climate a commanding officer establishes for the Marines in his or her charge. There is a disturbingly frequent correlation between Marines who act poorly and units with poor climates."[269] General Amos made a clear connection between the climate set by the commander and its influence on the behavior of the Marines in the organization. He underscored the importance of the commander setting the right climate and highlighted that the commander is the person responsible for that task: "Whoever's in charge is solely responsible for the command climate inside the organization, and that command climate will dictate the performance of the unit."

269. Amos, J. "Command Climate.". White Letter No. 3-13. Headquarters Marine Corps, Washington, D.C., 2013a, p. 1.

In 2012, General Amos responded to a series of sexual assault cases in the Marine Corps with a letter to all commanding officers and senior leaders in which he stated: "I expect each of you to do your duty and to set a command climate such that sexual assault is known among your Marines as shameful and unacceptable."[270] General Amos directed all commanders to set the right climate, which would push Marines to keep their moral compass, or, as he put it, their "*One True North.*"[271]

To further underscore the significance of climate and good leadership, General Amos discussed the frequency with which he covered command climate during his travels to Marine bases and stations: "I probably spoke more about command climate in my last two years than I ever recall any Commandant ever talking about it." Clearly, General Amos believes in the significance of the climate created by the commander as a critical aspect of good leadership, and instrumental in promoting ethical behavior. As part of the command climate, General Amos highlighted the value of open communication, dignity, and respect.

He suggested that open communication is one of the key policies the commander must implement when establishing the climate of the organization: "Number one, there had to be a constant line of open communication, absolutely, and it starts with the commanding officer and the Sergeant Major." He mentioned the need for a free flow of information, up and down the chain of command.

An example of open communication and how it promotes trust within an organization was when General Amos visited Marines in Iraq as a three-star general. He went to Hurricane Point, a Marine infantry command post in Iraq that was situated alongside the Euphrates River. The commanding officer greeted him and introduced him to one of his corporals (a junior NCO) who had been wounded in a firefight. The

270. Amos, J. "Leadership." White Letter No. 3-12. Headquarters Marine Corps, Washington, D.C., 2012b, p. 2.
271. *Ibid*, p. 1.

commander had the NCO escort General Amos around the battalion area. This demonstrated the complete trust and confidence that the commander had in that NCO. The commander knew the Marines by their first names, where they were from, and details about their families, all of which impressed General Amos. This example depicts a healthy organization, in which the commander is setting the right climate by trusting a junior Marine to escort and brief a very senior leader.

According to General Amos, "Leaders at every level are responsible to create an environment and command climate in which every Marine is treated with dignity and respect, one which enables Marines to trust their command, to take appropriate action."[272] General Amos discussed the importance of a command climate that upholds a genuine respect for one another in influencing the ethical behavior of Marines. For example, even when a commander must punish a Marine for violating one of the Corps' standards, it is done in a fair, compassionate, and professional manner, in which the Marine is held accountable (another key aspect of climate) but still maintains his or her dignity: "You take someone's dignity away, and you probably have either made an enemy for life, or whatever positive thing you were trying to do has just been done away with."

General Amos believes that you can get a good sense for the command climate of an organization in the way people treat one another, and how they talk to each other. Using the same example mentioned in the previous section, when General Amos visited the infantry unit in Iraq at Hurricane Point, he noticed the mutual respect the Marines had for each other and the deep affection they seemed to have for their commander: "Those units that have a good command climate generally always have those kinds of things in common—respect, dignity, open communication, and that kind of thing."

272. Amos, J. "Command Climate.". White Letter No. 3-13. Headquarters Marine Corps, Washington, D.C., 2013a, p. 2.

When he was asked if incidents of misconduct and poor ethical behavior would be reduced because of the right command climate, General Amos responded "yes, the answer is yes." For example, when he was the commanding general of the Second Marine Expeditionary Force at Camp Lejeune, he instituted a unit award called the "Chesty Puller Award" for great leadership. The criteria for the award were based on the number of reenlistments, non-judicial punishments, punitive actions, and successful enlistments (i.e., Honorable Discharges). General Amos recalled how he gave out these awards at various levels of command over his two years as the commanding general, and he noticed fewer discipline issues, traffic violations and vehicle accidents, and an increase in morale and respect. General Amos referred to General James Mattis (retired Marine Corps General and former Secretary of Defense), who used to say, "they're fighting with a happy heart."

Core Values and the Spiritual Health of the Corps

General Amos emphasized the Marine Corps' core values and the spiritual health of the Corps, as a commander's responsibility to reinforce to all Marines on a regular basis: "These core values have been the compass for every Marine's service throughout our rich history."[273] The spiritual health of the Corps is the "soul" of the Marine Corps. It is what is in a Marine's blood. It is the Corps' ethos. It is the sense of love for one another, the pride and devotion someone feels because they are a Marine, and that they would never do anything to tarnish the image of the Corps. As General Amos put it, "It is a religious thing, I think. And that's the thing that always must be garnered, cherished, nurtured, and carefully tended to by the commanding officers." The above statement underscores the importance of command climate and its influence on the ethical behavior of the Marines in the organization.

In 2012, General Amos delivered the Heritage and Values brief to remind Marines what they stand for, who they are, and who they are not. General Amos recalled how General Carl E. Mundy Jr. (30th Commandant)

273. DON, HQMC, *35th Commandant's Planning Guidance*, 2010.

told him that the institution is under attack from within (e.g., Marines violating the high standards of the Corps) and from the outside (e.g., Congress). According to General Amos, General Mundy always called him Jim, but when he mentioned the spiritual health of the Corps, he addressed him as General Amos and told him that he must not allow the spiritual health of the Marine Corps to decline, which could cause the American people to lose trust in the organization.[274]

Throughout the Heritage Brief, General Amos discussed scenarios where Marines violated the Corps' high standards and its core values of honor, courage, and commitment. For example, the hazing and suicide incident of a Marine in Afghanistan, or the Marines who urinated on dead Taliban soldiers, and the Marines holding up the flag of the Nazi SS in Afghanistan.[275] General Amos believes these examples were indicative of breakdowns in small unit leadership, accountability, discipline, and adherence to standards.

He brought up a meeting with the Joint Chiefs of Staff in the "Tank" at the Pentagon in 2012, during which the Chairman of the Joint Chiefs of Staff, General Martin Dempsey, said "those most responsible for maintaining our standards and discipline are allowing it to erode."[276] The Chairman reminded the Service Chiefs that it is their job to maintain the high standards and discipline within each of their respective Services.

In addition to the Heritage and Values brief, General Amos sent a letter to all of his commanders and senior leaders reinforcing the Heritage and Values brief and reminding them to get back to the basics of discipline, holding people accountable, and adherence to high standards. The letter stated that "a number of recent, widely-publicized incidents have brought discredit on the Marine Corps and reverberated at the strategic level. These incidents threaten to overshadow all our good work, and sacrifice. We are allowing our standards to erode."[277] These incidents

274. Amos, J. "Heritage and Values Brief [Power Point Slides]." 2012c.
275. *Ibid.*
276. *Ibid.*
277. Amos, J. "Leadership and Conduct." White Letter No. 1-12. Headquarters Marine Corps, Washington, D.C., 2012a, p 1.

point to the significance of a command climate where commanders lead by example, properly supervise, instill discipline, hold Marines accountable, and constantly reinforce the core values of honor, courage, and commitment. General Amos pointed to the NCOs as the next group of leaders who received his commander's intent and guidance, which he called "The Reawakening."

General Amos discussed the criticality of NCO leadership in assisting the commanding officers with the implementation of their command climate. The impetus for his urge to address every NCO in the Marine Corps face-to-face was the fact that out of approximately,174,000 enlisted Marines in the Corps, 144,570 were below the rank of sergeant, with 83% of the enlisted force led by NCOs. He stated: "Through your presence, leadership, and conduct, we will turn the tide of this battle against the insurgency of wrongdoing, restore our integrity with the American people, and keep our honor clean."[278] Clearly, General Amos was concerned about the spiritual health of the Corps when he said to the NCOs, "Move to the decisive point in this battle and through your presence, professionalism and tenacity…turn the tide of this fight for the sake of Corps and country."[279]

General Amos strongly believes that commanders must include beliefs in their guidance to Marines. He suggested that the quote from General Victor "Brute" Krulak[280] was the main message to all Marines and that the hinge point of the future of the Marine Corps centered on that quote and the climate that commanders establish in their commands. General Amos concluded: "We can't fade from ethical behavior and our core values. The Corps can't allow itself to become like everyone else."

Summary

In General Amos' view, the most important thing a commanding officer can do is to establish the right command climate that promotes ethical behavior within the organization. A good climate will result in an

278. Amos, J. The Reawakening. Headquarters Marine Corps, Washington, D.C., 2013b, p. 1.
279. *Ibid.*
280. See Krulak section.

increase in the number of Marines who abide by the Marine Corps' core values of honor, courage, and commitment. To facilitate the right climate, commanding officers must create an environment that encourages open communication, where people are able to establish close relationships with one another, built on trust and confidence in their abilities, as well as the trust and confidence of their peers and leaders. General Amos asserted that this type of environment promotes dignity and respect, where Marines treat each other according to the "Golden Rule."

He emphasized the importance of the Marine Corps' core values, and he discussed the spiritual health of the Marine Corps during the Heritage and Values brief. In that brief he spoke to all Marines about their lineage and how they are perceived in the eyes of the American public. He went on to publish a series of White Letters outlining his guidance to commanding officers on topics such as sexual assault, leadership, and command climate, underscoring his emphasis on values, and leadership. Further, he initiated The Reawakening Campaign, which was directed at NCO leadership. Here, General Amos asked the NCOs of the Corps for their leadership in assisting him to get the organization back on track "for the sake of Corps and country."[281]

281. Amos, J. The Reawakening. Headquarters Marine Corps, Washington, D.C., 2013b, p. 1.

General James Amos.

GENERAL JOSEPH F. DUNFORD JR., 36TH COMMANDANT (2014-2015)

Background

General Joseph F. Dunford spent roughly one year as the Commandant before becoming the next Chairman of the Joint Chiefs of Staff on October 1, 2015.[282] During this brief period as Commandant, Marines continued to act as America's first-response force, by deploying to crises around the globe supporting the combatant commanders. The Corps aimed to build forward-deployed and forward-engaged forces able to respond to contingencies, build partner capacity, build alliances, and project U.S. influence abroad.[283]

Marines continued to contribute to the mission in Afghanistan by supporting the Afghanistan National Security Forces against terrorism in Southwest Asia. Thanks to a heavy Marine Corps presence in the region, successful elections were held in the summer of 2014. Marines were able to transition the security mission to the Afghans. Marines continued to provide limited support to NATO in Afghanistan thereafter.[284]

In 2014, multiple Marine Expeditionary Units deployed in support of the geographic combatant commanders, providing forces in support of their areas of responsibility. For example, in 2014, Marines conducted operations in support of the U.S. Embassy in Libya.[285] Marines continued to deploy around the world, providing reassurance to America's allies and acting as a deterrent to potential foes. In addition to security operations in support of American diplomats, Special Purpose Marine Air-Ground Task Force Crisis Response units conducted various missions such as embassy reinforcement, training exercises, and combat operations against the Islamic State.

282. See Appendix E.
283. Dunford, J. Brief to the House Appropriations Committee Subcommittee on Defense. Headquarters Marine Corps, Washington, D.C., 2015a.
284. *Ibid.*
285. *Ibid.*

In the Pacific region, Marines conducted numerous operations and exercises such as Exercise Song Yong—the largest amphibious exercise of the year, in partnership with the South Korean troops. The Corps positioned approximately 22,500 Marines west of the International Date Line to act as forward-deployed forces in the Asia-Pacific Theater. During the next three decades, Marines will be positioned in Australia, Guam, Japan, and Hawaii. This plan enables Marines to deploy to Australia, conducting bi-lateral training and exercises with the Australian Marines, and will be postured to support other military operations such as humanitarian assistance/disaster relief.[286]

In addition, Marines deployed in support of security and stability operations in Europe, an example being the Black Sea Rotational Force mission, which conducted numerous activities in Europe and provided Marines to act as a crisis response force. An example of this is the Fleet Anti-Terrorism Security Teams, which has supported embassy reinforcement missions in Baghdad, Iraq and Sana'a, Yemen. Marines continued to support the State Department with Marine Security Guards, manning 173 embassies and consulates in 141 countries around the globe.[287]

The Interview

General Dunford was interviewed in his office at the Pentagon in December 2017, for approximately one hour. The primary message from General Dunford was that the commander is responsible for the "intangibles" of the unit. The intangibles he discussed were areas such as character development and being able to communicate to Marines the importance of the core values, and how they influence the Marine Corps' warfighting effectiveness and readiness.

286. *Ibid.*
287. *Ibid.*

This section discusses command climate and the intangibles of core values, and trust in small unit leadership. In addition, the section addresses how to establish the right climate through setting the example, implementing the concept of balanced excellence, establishing high ethical standards, and being a good communicator-in-chief.

The Intangibles of Command Climate

General Dunford reflected on an incident when he was a young captain that had a profound impact on his approach to command climate and the value of the intangible qualities a leader can bring to the organization. He was attending a formal dinner (mess night) in Fuji, Japan and the officers in attendance had been drinking too much, which caused the night to unravel. Brigadier General Hank Stackpole was the guest speaker and stood up before his scheduled remarks. Once he had everyone's attention, he proceeded to speak about what it meant to be a Marine officer. Then, he summarily dismissed the evening and sent everyone home before the scheduled conclusion. "It stuck with me," recalled General Dunford, "I don't know why, but every time I see him, I relay that story and what an impact it had on me as a young officer."

General Dunford believes this act required a significant amount of moral courage by General Stackpole. It demonstrates that it is the responsibility of the senior leader, or the commanding officer, to ensure that Marines behave according to Marine Corps values and standards. This example by a senior leader exhibits the intangible nature of moral courage, which is one of the core values.

When asked if command climate influences ethical behavior, General Dunford responded: "I think it is the primary driver of ethical behavior. It all comes from the commander." For General Dunford, the commander establishes the right climate within which the Marine Corps' core values are taught and reinforced. He emphasized to those attending the course for new commanders the significance of establishing the proper climate in which the Marine Corps' values are maintained, from the time a Marine completes boot camp and joins their new organization. As part of his message to the new commanders, General Dunford pointed to who Marines are and what the American public expects from its Marine Corps.

He suggested that the only way America will defeat organizations such as ISIS, is through our American values: "When we go to war, for example, we bring our values with us. The issue for us in the long term is can the United States after all these years at war maintain the ethical standards that define us." General Dunford's message highlighted the significance of core values training and regular reinforcement of that training.

Also, General Dunford discussed how leaders can institutionalize command climate and ethical behavior: "If you say or do something that is inconsistent with our core values it is not a reflection on you, it is a reflection on all of us." He mentioned the sense of accountability Marines have for one another, and when they do something that contradicts the core values, they are letting down the Marine on their left and the Marine on their right.

General Dunford believes there is a general decency in the young people entering the Marine Corps today, who want to be part of something bigger than themselves: "They aspire to those core values that the Corps advertises. So, you have a pretty good group coming in." To continue to reinforce the Marine Corps' core values, small unit leadership is required to assist the commander with the character development of the Marines in the organization.

General Dunford suggested that NCOs are critical to the ethical health of the Marine Corps and instrumental in sustaining the Transformation of becoming a Marine as a quality citizen. The general quoted a Marine Corps Reference Publication[288] in his planning guidance: "Significant to

288. MCRP 6-11D, *Sustaining the Transformation*, 2014.

sustaining the Transformation is selecting the best Marines to be NCOs and continuing to train them to sustain the values and warfighting ethos of our Corps."[289] He used a historical example to further highlight the significance of small unit leadership and sustaining the core values that contribute to warfighting effectiveness. He suggested that I read *With the Old Breed* by E.B. Sledge.

Sledge was a World War II Marine who fought with the First Marine Division at Peleliu and on Okinawa. Sledge attributed his ability to survive months of horrific combat against the Japanese to his trust in the small unit leadership of the junior officers and the NCOs.[290] General Dunford discussed ethical behavior with trust as a subset of that behavior and referenced the intangibles (e.g., esprit, trust, ethical behavior, respect, selflessness) associated with small unit NCO leadership as depicted in *With the Old Breed.* As Sledge put it, "It was trust. It was trust in my leadership, trust in myself because of my training, and trust in the Marine on my left and right.[291]

Dunford said, "That is the commander's business." In other words, the commanding officer is responsible for setting the right climate that promotes the intangible qualities that identify Marines and requires the support of the small unit leadership (junior officers and NCOs) to reinforce those intangibles as part of the commands' climate.

Establishing the Right Climate

General Dunford addressed three major areas critical to establishing the right command climate: setting the example, establishing high standards, and being able to effectively communicate those standards to Marines: "Command climate starts with the example of the leader, the role model,

289. Dunford, J. Brief to the House Appropriations Committee Subcommittee on Defense. Headquarters Marine Corps, Washington, D.C., 2015a, p. 6.
290. Sledge, E.B. With the Old Breed. New York, NY: Presidio Press, 2007.
291. *Ibid.*

the individual who articulates what ethical behavior and what standards exist. It all comes from the commander."

General Dunford argued that "the best way to communicate ethical behavior is by personal example, and then by the priorities you establish as a commander." The general suggested that commanders must be visible and that they must get around to see how their Marines are doing. By doing so, commanders can verify that their Marines have the means necessary to accomplish the tasks that they are responsible for within the unit. Commanders must also ensure that the subordinate leaders are taking good care of the Marines. For example, commanders must ensure that Marines have adequate equipment, that they are being accurately paid, and that they are receiving their awards on time. As General Dunford put it, "it is what you do as opposed to what you say." In other words, commanders must live the example—walk the talk—and not just talk about what they plan to do.

Additionally, General Dunford emphasized the personality of the commander relative to setting the right climate and being able to influence ethical behavior. Commanders cannot have egos that get in the way of their ability to lead their Marines. In regard to leaving your ego at the door, General Dunford said, "I think it is a personal example more than anything else." He discussed famous Marines who were paternalistic figures, such as Chesty Puller, who had five Navy Cross decorations (the award second only to the Medal of Honor) and was a humble commanding officer, and O.P. Smith, who commanded the first Marine Division during the Korean War at the Chosin Reservoir, but always downplayed his incredible accomplishments giving the credit to his subordinate leaders and to his beloved Marines.

These men epitomized the idea of service to others before service to self and left their ego at the door. General Dunford suggested that selflessness and serving the needs of your Marines first is the best way to set the example when communicating with 19-22-year-old Marines. Though sometimes this might get in the way of other command prioirities, it takes balance to ensure the commander covers everything that needs to be taken care of for the overall health of the organization.

General Dunford recommended a philosophy of command built on balanced excellence. The general indicated that the commanding officer is responsible for everything the unit does or fails to do, including training, maintenance, and personnel readiness. Yet, many of these tasks can be accomplished by subordinate leaders.

A unit must be proficient in its mission essential task list, but at the same time complete other required annual training such as troop information classes, swim qualifications, individual career counseling, sexual assault awareness and prevention, rifle requalification, and so on. According to General Dunford, the commander is directly responsible for the intangibles discussed earlier, such as ethical behavior, trust, core values, respect, and commitment:

> On the one hand, you have to develop proficiency in your mission essential tasks. On the other hand, you have to take care of all the equipment and maintenance. But then in the middle is what the commander is responsible for. And that is all the intangible qualities.

General Dunford also referenced the Marine Corps publication *Leading Marines,* where he believed some of the intangibles are discussed that help guide commanders on areas relative to the standards that are expected within the organization.[292] He directed all Marines to be committed to the Marine Corps' values and its high standards: "I also need the full commitment of all Marines to our high standards of professionalism, discipline, and core values."[293] General Dunford believes it is

292. DON, HQMC, 2014, MCWP 6-11, *Leading Marines*.
293. Dunford, J. Brief to the House Appropriations Committee Subcommittee on Defense. Headquarters Marine Corps, Washington, D.C., 2015a, p. 3.

the commander's responsibility to create the environment that promotes ethical behavior. As such, how does the commander respond to incidents of misconduct? In other words, how does the institution respond by holding people accountable in the wake of an incident?

General Dunford highlighted the importance of Marine Corps values and how Marines must abide by their high standards, 24/7, 365, including combat operations. He underscored the need for ethical behavior in combat through discrimination, proportionality, and sound judgment, for instance regarding civilian casualties.

When he was Commandant, and as the current Chairman of the Joint Chiefs of Staff, the guidance General Dunford published on the rules of engagement are a part of his command climate:

> The Secretary of Defense and I are rewriting all of our instructions that relate to civilian casualties in humanistic language that talks about respect for human life and why it's important that we conduct operations in a way that absolutely minimizes the loss of innocent human life.

General Dunford recalled his command tour of duty in Iraq as a regimental commander preparing to move his Marines across the Diyala River. As a colonel and in command of a large group of Marines, he accepted an increased risk to his force when he decided not to shoot at enemy targets on the far side of the river, due to the potential for innocent civilian casualties. "Respect for human life is our business," he declared, "I also think it mitigates the risk of moral injury on the backside of a conflict." Marines value life and are expected to live up to the Corps' high ethical standards, especially during combat operations. This enables them to maintain the moral high ground and be able to live with their decisions and actions once they return home.

General Dunford mentioned that now is not the time for our commanders to be the strong, silent type of leaders. Today's leaders must be able to effectively communicate their expectations of behavior and their expectations of ethical conduct to their subordinates: "Your personal example has to be accompanied by talking to people and I think using the tools that are available to us."

He mentioned the tools commanders have available to them to reinforce ethical decision making and how to handle ethical dilemmas such as modern immersive training and simulation technologies.[294] These technologies allow the Marine to develop the ethical "muscle memory" necessary to prepare the Marine for the pressures associated with ethical decisions under extreme conditions.

He also discussed the value of using the Socratic method as a technique that can generate dialogue causing the Marines to examine the facts of the scenario and discuss their decision making process: "The 'what if' you're confronted with this kind of a situation and a conversation about how to deal with that is, in my judgment, one of the more important ways to teach people."

Finally, General Dunford discussed the significance of communicating the sense of obligation each Marine has to the institution:

> So, from the earliest days in recruit training, you are instilling in a private, the sense of accountability to the values of the institution and the sense that if you violate those values, it's not about you. You are actually doing something that is putting a black mark on the institution, as a whole, and that is particularly true the more senior you become.

When General Dunford was Commandant, he would speak to the new general officers, and he would tell them that they no longer have

294. *Ibid.*

personal opinions when speaking in public: "You can never again speak in public without being a reflection of the institution. And when you are communicating publicly, you actually don't have a personal opinion because you are the institution." What generals discuss or do in public represents the position of the Marine Corps and reflects on the Marine Corps as an institution. To underscore the significance of the command climate and the importance of winning in combat while maintaining the Marine Corps' core values, General Dunford asserted: "Although we remain proud of our heritage, we should expect no credit tomorrow for what we did yesterday."[295]

Summary

General Dunford discussed the intangibles of command climate (e.g., trust, values, respect) as the responsibility of the commanding officer. He emphasized that commanders must create a climate that sustains the core values training that Marines receive during entry-level training (e.g., boot camp or Officer Candidate School). For commanders to successfully accomplish this task, they need to rely on their junior officers and NCOs. This promotes trust within the organization and enables junior leaders to reinforce the command climate.

General Dunford identified setting the example as the first critical element for a commander who desires to establish the right command climate. Commanders are expected to do what they say they will do and to take good care of their people. Commanders have many competing priorities in the training schedule and sometimes core values training and reinforcement does not receive the attention it deserves. Therefore, he recommended a balanced approach to command climate, where commanders focus on the intangibles and let their staff and subordinate leaders take care of many of the other tasks, such as maintenance of equipment, the training schedule, and personnel matters.

295. *Ibid*, p. 5.

Finally, General Dunford suggested that the commanding officer must be a great communicator-in-chief, able to talk to Marines about what their service means to the country. It is the responsibility of the commander to teach and reinforce the core values as part of the command climate. Commanders must be able to communicate to their Marines that it is their responsibility to never do anything to tarnish the image of the Marine Corps, which could cause the American public to lose trust and confidence in its Marines.

General Joseph Dunford and the author.

GENERAL ROBERT B. NELLER, 37TH COMMANDANT (2015-2019)

Background

General Neller became the 37th Commandant of the Marine Corps on September 24, 2015.[296] At the time, Marines were forward deployed in Iraq and Afghanistan and responding to crises around the world. Since then, Marines have been involved in numerous types of operations around the world such as building partner capacity, theater security cooperation efforts, training and working with allied partners, and preparing for future contingencies. For example, in 2015, Marines executed approximately 100 operations, 20 amphibious operations, 140 theater security cooperation events, and 160 major exercises. Today, Marines are serving at 174 Embassies and Consulates in 146 countries around the world.[297]

The Marine Corps has deployed thousands of Marines on Navy amphibious shipping to all areas of the globe. Recently, five separate Marine Expeditionary Units supported every geographic combatant command participating in training exercises. In addition, Marines have been involved in disaster relief operations, such as in Saipan following Typhoon Soudelor, where Marines provided 11,000 gallons of fresh water and 48,000 meals within 12 hours after notification of the request for support.[298] Marines were also deployed to the U.S. Southern Command, tasked with the reconstruction of a runway at Mocoron Airbase, Honduras.[299]

As part of Pacific Command's forward-deployed force for crisis response, 22,500 Marines west of the International Date Line operate within the Asia-Pacific Theater. Also, approximately 1,200 Marines deployed to Darwin, Australia for a six-month deployment.[300] General Neller stated

296. See Appendix E.
297. Neller, B. *The Commandant's Posture of the United States Marine Corps, President's Budget* 2017.
298. *Ibid.*
299. *Ibid.*
300. *Ibid.*

to Congress that the Corps will continue to produce highly-trained Marines that are formed into combat-ready forces, who can provide the capabilities that the Joint Force requires.[301]

The Interview

General Neller was interviewed in his office at the Pentagon, on November 21, 2017, for approximately one hour. During the interview, the general emphasized that the right command climate is an outcome of the commander's leadership, which sets the conditions for Marines to be successful and to become better people. In other words, commanders set the right climate for their organizations and are responsible for developing their Marines into better people, which will produce better Marines. The Commandant centered his remarks on the philosophy of command, and the commander's approach to leading people, which includes holding them accountable for their actions and teaching them to be responsible citizens.

This section addresses the philosophy of command. General Neller outlined how commanders must communicate their ethical message to the Marines, along with the commander's expectations for the Marines in the organization. The commander is expected to set the proper conditions for individual and organizational success. Finally, the section addresses leading people, particularly setting the example, servant leadership, effective communication, accountability, and responsibility.

Philosophy of Command

"Commanders have an obligation," General Neller began, "to do all they can to make all their Marines and Sailors successful." This obligation entails creating an environment where Marines can achieve their individual goals if they are willing to work hard and if they abide by the

301. *Ibid.*

Marine Corps' core values of honor, courage, and commitment. Two of the key elements of the philosophy of command are the expectations for both the leader and the followers, and that it is the commander's responsibility to set Marines up for success.

Commanders must ensure that Marines understand who they are, where they come from, and why they are here: "But when you're a Marine, you are expected to follow orders. You're expected to be disciplined." Marines must understand that they took an oath to serve, with no expiration date attached to it.[302] In other words, Marines are expected to act like Marines 24/7, 365 days a year, and understand that they are Marines for life. The command philosophy provides the commanders an opportunity to cover how they like to do business, and how they plan to approach the mission. Commanders can also discuss how they like to communicate and what they want to accomplish during their time in command. The philosophy of command is also a good instrument for the commander to tell the Marines what they can expect from their commander. According to General Neller, "if you tell Marines what you expect from them, you should tell them what they should expect from you."

The Corps puts a lot of pressure on its commanding officers, maybe too much pressure, according to General Neller. However, commanders in the Marine Corps are held to an even higher standard than the Marines they lead, which can cause them to be judged unfairly. But commanders are still expected to set the tone for their organizations while demonstrating that they are men and women of the utmost character. General Neller emphasized this point: "They expect us to hold people accountable, ourselves first and foremost, and to be somebody of virtue and character."

---

302. Neller, B. "A Message from the Commandant of the Marine Corps." 241st Birthday of the Marine Corps. Headquarters Marine Corps, Washington, D.C, 2016.

General Neller discussed the importance of treating people fairly and how that relates to the ethical behavior of Marines. He mentioned that Marines keep track of who gets punished and who does not: "Marines, like your kids, they all keep score. So, fairness can be relative." Similarly, General Neller referenced the book *Legacy*, which addresses the value of treating people fairly with no tolerance for "dickheads."[303]

Kerr suggests that it is the leaders' job to promote trust and collaboration, where "success can be traced back to the connections between members of the team and their collective character, something true of all winning organizations."[304] Clearly, the commanding officer must meet many expectations that include setting the right conditions for the Marines in his or her organization to succeed.

General Neller suggested that many young people come into the Marine Corps from non-traditional homes: "When 40% of young men and women grow up in this country without necessarily, what we would call a normal family, whatever that is, I'm not sure everybody knows what the right thing is. So, it's up to us to teach them." One of the first steps to setting Marines up for success is to ensure they understand the Marine Corps' values and that they meet their commander's expectations and the expectations of their fellow Marines.

Also, the commander must create an environment that prides itself on the character of its Marines, but also promotes initiative and growth through learning in an environment that is not based on a "zero defects" mentality. General Neller offered an example of a good learning environment and of a command climate that seemed to be working. During one of his recent trips to Twenty-nine Palms, California, he spent the night with one of the infantry battalions training in the field. In the morning,

303. Kerr, J. *Legacy: What the All Blacks Can Teach Us about Life.* London, UK: Constable and Robinson Ltd, 2013.
304. *Ibid*, p. 83.

he listened and observed how the unit leaders moved the Marines: "My sense is that the unit that did it with the least amount of noise, the least amount of yelling, was probably the more effective unit." He suggested that the better units had junior Marines communicating and checking on the troops, exercising initiative, while the senior leaders were just monitoring what was going on that morning.

As part of his guidance to all Marines that will set the conditions necessary for them to succeed, General Neller advised, "Drink less. That's all. I'm not telling you not to drink. You're a grown man. Drink less. Or stop. Because I can guarantee you, it'll improve your prospects for success in life."[305] He suggested that one of the biggest issues in the Marine Corps is that a small percentage of Marines abuse alcohol, which causes many of the misconduct problems in the Corps. For example, he mentioned a recent incident in Okinawa where two Marines and a Sailor, out of approximately, 22,000 Marines, did something that resulted in prison time and bad characterizations of service. These incidents of misconduct reflect poorly on the Marine Corps as an institution, causing the Japanese and other allies in the Pacific Rim to form a negative opinion of the Marine Corps. It is the responsibility of the commander to ensure the Marines understand that they own the blank space associated with their reputation and that they are well educated on the core values. For example, a Marine must protect his or her reputation, and each Marine owns the space after their name. As General Neller put it, "you won that space and what's it going to say? What do you want it to say? Because, it'll say whatever you decide it's going to say, good, bad, or otherwise." Therefore, it is the commander's responsibility to ensure he or she establishes a command climate that promotes the necessary behavior for Marines to be successful, and exludes things like the abuse of alcohol that may lead to a path of destruction.

305. Neller, R. "Learning Without Losing." The *Marine Corps Gazette* 101 no. 11 (2017a): p. 7.

Leading People

General Neller discussed the atmosphere when he joined the Marine Corps during the 1970s, when, he believes, the Corps was not very good. He observed various leaders who took advantage of their positions, causing some Marines to lose confidence in the chain of command. However, General Neller stayed in the Corps: "I bought into the idea that if you can convince a bunch of people to be like-minded about a mission which is worthy and noble, and you could convince them it was important, and you could do that through your own behavior and actions, then that was kind of a cool thing to do." He went on to reflect on the importance of setting the example, selflessness, and effective communication.

General Neller emphasized the significance of setting the example and emphasized that it is the commander's responsibility to model the expected behavior: "If their leaders that are teaching them don't model it, it's either seen as simply words or hypocrisy." The general also suggested that units who are not effective have leaders who do not do what they say they will do and who do not model the right behavior. "Marines watch everything you do," he commented, "and hear everything you say. You are on parade all the time." Leaders are expected to set a good example for others to follow and to be a man or woman of character. They must take care of their Marines and focus on the needs of the Marines before they take care of themselves. Commanding officers must not take advantage of their positions or be perceived as having an advantage because of their rank or status: "Sometimes people get in a position of authority as commanders, and they can lose their way." Sometimes, commanders let the little things go that are very minor violations of rules and regulations. Over time, this can lead to bigger problems, especially when their ego gets involved.

General Neller suggested that, as a commander, it is not about you and that commanders should focus on the needs of their Marines before their own needs are met. One example he used was from when he was a captain and his infantry unit was conducting cold weather training at Bridgeport, California. The Marines had been in the field for days. One night he arranged for his Marines to eat hot chow in the field. When the food arrived, one of his junior officers grabbed some cake from the truck. One of General Neller's Marines observed this act and complained to him that the lieutenant had cake before any of the other Marines. This example underscores the negative perceptions that are associated with leaders whenever they take care of themselves before taking care of their subordinates. General Neller had to address this with his lieutenant.

Another example he shared underscored the importance of commanders leaving their egos at the door. He believes that General Oliver P. Smith, hero from the Korean War, epitomizes selflessness and humility: "Commanders need to check their ego at the door. O.P. Smith is a great example." He mentioned how Smith always put the needs of his Marines first by sharing the risks with his Marines during battlefield circulation visits, serving alongside his men in combat, wearing the same cold weather gear that they wore in forty-degrees-below-zero conditions, eating the same cold food from cans, and questioning senior leadership when they were wrong.[306] General Neller believes that Marines join to serve the Marine Corps: "They're here to serve not us, they're here to serve the Marine Corps." It is the job of the commanding officer to take care of the Marines and make sure they have their needs met before the commander takes care of himself. This is another area that will promote good ethical behavior and produce better Marines.

One of General Neller's concerns dealt with how Marines communicate: "I just worry about communication, how we communicate." Today's generation of Marines rely on social media. The Commandant expressed concern that Marines are losing the skills of having face-to-face conversations. As Commandant, General Neller directed all leaders to address

306. Shisler, G. *For Country and Corps, The Life of General Oliver P. Smith*. Annapolis, MD: Naval Institute Press, 2009.

the expectations and conduct associated with the use of social media.[307] He directed mandatory counseling for all Marines regarding the use of social media, where he emphasized core values, social media education, and leadership from commanding officers.[308] He did, however, mention that social media is a double-edged sword, as it provides leaders with another way to pass information: "Social media is a very powerful tool for a commander to communicate their thoughts and ideas to address perceptions, or to reinforce their ideas, concepts, and philosophy."

He also brought up his concerns with online training. He believes that the Corps needs to get back to small group discussions led by NCOs and junior leaders, who can communicate current issues facing Marines and to assist them in developing solutions to the problems. Although he pointed out that online training is better than no training, he emphasized the value of personal relationships and face-to-face communication with instructors as the preferred option.

Accountability and Responsibility

General Neller recommended for commanders to hold their Marines accountable. For their part, the commander should be held to an even higher level of accountability: "He has got to hold himself more accountable than anybody else in the unit." He also highlighted that it is the commander's responsibility to establish an environment where Marines take responsibility for their actions.

Marines expect their leaders to hold them accountable: "They expect us to hold people accountable, ourselves first, and foremost." General Neller mentioned that most of the commanding officers are relieved because of misconduct, not incompetence. Therefore, it is incumbent

307. Neller, B. "Recent Conduct on Social Media." White Letter No. 1-17. Headquarters Marine Corps, Washington, D.C, 2017b.
308. Neller, B. "Social Media Guidance—Mandatory Counseling Requirement." White Letter No. 2-17. Headquarters Marine Corps, Washington, D.C, 2017c.

upon commanding officers to hold themselves accountable for their actions. Their character should be without reproach.

The Commandant also discussed the importance of holding Marines accountable when they make a mistake of commission as opposed to a mistake of omission. In other words, commanders must hold their Marines accountable when they violate rules, regulations, and Marine Corps standards, but they need to create a command climate that promotes initiative and enables Marines to make mistakes. This environment contributes to setting Marines up for success, so they can accomplish their goals. If the commander is unsure as to whether a violation of standards has occurred, then it is that commander's responsibility to investigate the matter to obtain all the facts and determine how best to hold the Marines involved accountable.

General Neller raised the question, "Are we a self-correcting organization?" Commanders hold their Marines accountable. Marines expect their leaders to be held accountable. Does the Marine Corps as an institution hold itself accountable? General Neller suggested that the Corps must always strive for improvement, so that it can be the most capable and the most lethal warfighting organization in the world. To that end, Marines must be willing to accept criticism and make the necessary adjustments to continue to improve their efficiency and effectiveness as a military organization. This includes Marine commanding officers.

As officers are trained, the Corps instills in them the conviction that the commanding officer is responsible for everything his or her unit does or fails to do.[309] General Neller, for example, charged his commanders with carrying out his guidance after the social media scandal "Marines United." He said: "Leaders at all levels must address online sexual harassment, threats of violence, and other misconduct, as we would address

309. DON, HQMC, MCWP 6-11, *Leading Marines*, 2014.

any other form of misconduct or destructive behavior."[310] Commanders are responsible for everything that goes on in their organization, for example, training and education, maintenance of equipment, family, and personnel readiness.

Marines are expected to be responsible people, and as such, they are expected to follow orders and abide by the core values. General Neller acknowledged that Marines will make mistakes and that is also to be expected:

> I guess the real question is why is it so difficult for some to follow? Is it weakness of the human condition, the weakness of human character, or temptation, or the impact of alcohol or drugs? Is it a lack of humility when you assume a leadership position? Is it a character flaw? I don't know.

Ultimately, it is up to the commanding officer to ensure the Marines understand and acknowledge the standards. If they choose to be irresponsible, they will be held accountable.

Summary

General Neller believes that it is important to ensure the Marines understand the commander's expectations. It is incumbent upon the commander to let his or her Marines know what they can expect from him or her. General Neller suggested command philosophy as one way to convey these expectations to Marines. He underscored that the command philosophy and organizational climate must promote an environment in which Marines are set up for success, so they can accomplish their individual goals while in the Corps and in life.

He emphasized the importance of setting the example when leading others. He mentioned that it is the commander who must ensure the

310. Neller, B. "Recent Conduct on Social Media." White Letter No. 1-17. Headquarters Marine Corps, Washington, D.C, 2017b, p. 1.

needs of the Marines are addressed first, before his or her needs and the needs of the other officers. He voiced his concerns on how Marines communicate, and that social media is taking the place of face-to-face dialogue. He suggested that the Marine Corps return to personal communication and small group discussions as the best way to learn. Finally, General Neller suggested that the commanding officer hold himself or herself to a higher level of accountability. Commanders must hold their Marines accountable, and if they are not sure of the facts, then they should investigate. Also, Marines are expected to be responsible people even though they are not expected to be perfect. It is understood they will make mistakes. It is the commanding officer's task to make sure that the command climate includes training where the focus is on the individual Marine's character.

General Robert Neller and the author.

Photo by alancrosthwaite.

CHAPTER THREE

Cross-Case Comparison

A Marine Corps commanding officer is directly responsible for everything his or her organization does or fails to do. It is the responsibility of the commanding officer to create the right environment that promotes good order and discipline, while enabling the organization to accomplish its mission.[311] It is also the responsibility of the commanding officer to make Marines better people than before they came into the Marine Corps.[312]

The purpose of this study was to examine the role that command climate plays in influencing the ethical behavior within a Marine Corps command. This project's exploratory multiple-case study/cross-case analysis design, supported by qualitative research methods, examined eight living CMC, which included the current Commandant and the current Chairman of the Joints Chiefs of Staff, also a former Commandant.[313]

Specifically, this study systematically examined and analyzed the individual perspectives of eight CMC—covering three decades of the Marine Corps' most senior leaders and their perspectives about command climate and how it influences the ethical behavior of the Marines in the organization. Although the CMC provided their individual perspectives, four major categories emerged from the individual interviews. These were grouped for comparison as listed on the next page.

---

311. DON, HQMC, MCWP 6-11, *Leading Marines* 2014.
312. Lejeune, John A. *Reminiscences of a Marine*. University of Wisconsin-Madison, WI: Arno Press, 1930.
313. Merriam, S. and Tisdell, E. *Qualitative Research, A Guide to Design and Implementation*. San Francisco, CA: Jossey-Bass, 2016.

1. Setting the example
2. Open communication
3. Core values
4. Accountability and responsibility

Each Commandant offered his perspective on these categories, and how it contributes to the commander's climate and its ability to influence the ethical behavior of the Marines. Three major themes emerged out of the four categories discussed in this chapter. Theme One: The command climate set by the commanding officer is significant with respect to influencing the ethical behavior of the Marines in the organization. Theme Two: The commanding officer's climate must be focused on the Marine Corps' core values, and the character development of the individual Marine post entry-level training (sustaining the Transformation). Theme Three: The best way for the United States Marine Corps to continue winning the hearts and souls of the American people, which is directly related to the Marine Corps' survival as an institution, is through the ethical behavior of the individual Marine both on and off the battlefield.

Every Commandant had his own personal way of conveying his philosophy of command. Some emphasized certain areas over others. For example, General Krulak focused most of his points on the core values, while General Jones emphasized setting the example. All the CMCs, however, emphasized the importance of protecting the Marine Corps as an institution, which fell under the category of accountability and responsibility. Based on the interview data, it is the Commandant who is ultimately held accountable to ensure that the Marine Corps continues to remain America's most formidable warfighting organization, capable of fighting and winning the nation's battles.

Again, the primary concern that guided this research was: What role does command climate play in influencing the ethical behavior within a Marine Corps command?

This was examined through two research questions:

1. How do the CMC describe a command climate that encourages ethical behavior among Marines?

2. How does a commanding officer develop a command climate that promotes ethical behavior?

The CMC were studied as individual cases. Then the individual case studies were analyzed and compared. This section focuses on the four categories that were gleaned from the cross-case comparison beginning with setting the example. This discussion is followed by a section that provides an overview of the three themes produced from the four categories, which provides the foundation for the recommendations discussed in greater detail in chapter four. Finally, a cross-case comparison summary concludes the chapter.

Categories

1) Setting the Example

Each Commandant discussed the importance of setting the example. The CMC discussed how leaders set the example and that the example must start with the senior leader through his or her own personal example. Also, commanders must be visible, so the troops can witness the commander's example. Finally, to set the right example, the leader must be selfless.

General Dunford emphasized setting the example when he discussed the importance of the message coming from the senior leader: "Without proper command climate that starts with the example of the leader, the role model, the individual who articulates what ethical behavior and what standards exist for ethical behavior. It all comes from the commander." Generals Amos and Neller aligned with General Dunford when they both suggested that the commander sets the example and the command climate for the organization. As General Amos put it, "whoever's in charge is solely responsible for the command climate inside the organization, and that command climate will dictate the performance of the unit." General Neller agreed: "The commander's got to model the behavior that he expects every other Marine to exhibit." General Neller recommended that commanders set the example by doing what they said they promised to do: "We've got to do what we say we're going to do. And I think most units that are not effective don't. The leaders don't."

Generals Hagee and Neller both suggested that it is the little things that commanders do, or do not do. General Neller used the example of the junior officer who had a piece of cake before the Marines had a chance to eat: "It's the little things, it's a lot of little things." In this case, it was the idea that the officers eat after the troops.

Interestingly, General Krulak was one of the few generals who suggested that the Commandant is the one who sets the climate for the Marine Corps: "So, at the end of the day, what sets the climate, whether it's to a platoon or the Corps is one person, one person only. And that's the Commandant. And how he acts is critical, and what example he sets, and does he walk his talk." As General Gray said, "you have to live it, and you have to believe it." In other words, what the leader professes to the Marines, and the expectations he or she sets for them, must be modeled by the leader.

Leaders must truly believe in the high standards and the high expectations that he or she establishes for their Marines. General Gray said it most succinctly: "You have to set the example. But you must not only set it, you have to live it." He mentioned the significance of "living it" through your own actions and creating an environment where you get your Marines to follow your example because they see you doing it: "You need to do it in such a way that you're not acting like a guy on a big white horse. You're doing it because you believe in it. You're setting the example because you hope that other people will follow it." Generals Krulak and Amos both said, "you just live it, you live it."

Generals Krulak, Jones, Gray, Conway, and Neller all advocated for getting out and letting the Marines see your leadership by example. General Krulak discussed the time when he was a battalion commander and went to the unit's motor pool, donned a set of coveralls, and proceeded to check under the hoods of military vehicles for basic maintenance checks. The troops observed him working next to them doing the same job that they were doing. General Jones suggested that the commander must be visible and show the Marines how the leader acts: "I mean, you have to show people how you handle yourself ethically, you know, people need to see that."

General Gray changed into civilian clothes and walked around the Okinawan town of Kinville to talk to the Marines and the owners of the local bars to see how he could reduce the number of alcohol-related incidents by Marines. Although he was not a drinker, he wanted to take care of his Marines, and he eventually developed a courtesy patrol that brought Marines back to the barracks after they had consumed too much alcohol. This act reduced the number of alcohol problems in his command. General Jones likewise underscored the significance of being visible while setting the example: "How you live your life and how you are seen by others is very important."

General Krulak remained visible as he travelled over 700,000 miles while Commandant, more than double the miles travelled by any other CMC, so he could visit the Marines twice per year at every post and station in the Corps. This was one of his techniques of setting the example for his Marines by showing them he cared enough about them to travel around the world to see them and to share his Commandant's guidance with his Marines.

General Conway emphasized the importance of getting out and talking with the troops about his philosophy of command: "They've got to see you when you get intense and see you when your eye starts to twitch, and your brain starts to bulge, and those types of things, so they understand their commander." General Neller suggested that commanders must be visible and that they must set the example by sharing the risk with their Marines: "But you've got to show up. You've got to compete. They've got to see you out there sharing the risk, or the deprivation, or the pain."

Generals Jones and Neller both used the example that officers eat last when they suggested that the leader must take care of the followers first before taking care of himself or herself. General Jones' father was the youngest battalion commander in World War II at age 27, and role-modeled values and ethics. General Jones recalled that "they dominated our discussion, and we lived by it. Even when I was seven years old, I remember my father saying, 'Officers eat last.'" In other words, the needs of the officers come second to the needs of the troops. Similarly, Generals Dunford and Neller both discussed the importance of selflessness when leading others. As General Dunford put it, "service to others before service to self." General Neller discussed the reason why people join the Marine Corps and expressed his perspective on what the leaders must remember: "We're here to serve them. They're not here to serve us; they're here to serve the Marine Corps."

General Hagee also discussed the concept of selflessness and used Admiral Nimitz as his primary example. Admiral Nimitz was offered the position of Chief of Naval Operations by President Roosevelt, which Nimitz declined. Nimitz believed it would not be good for the Navy as a junior admiral to jump over so many other senior officers who were eligible for the position. As General Hagee said: "It was never, ever about him." Nimitz always put the needs of others, or the Navy as an institution, over his own needs and desires.

Like General Neller, General Dunford discussed the importance of setting the example: "It is kind of what you do as opposed to what you say." Again, every Commandant emphasized setting the example and that the command climate starts with the commander and the example that he or she provides for the others to follow. Another category where the CMC seemed to agree collectively was the need for the commander's climate to include open communication.

2) Open Communication

Each Commandant suggested that open communication is necessary for a healthy command climate and one that influences the ethical behavior of the Marines. Although General Krulak did not specifically use the words "open communication," he discussed the importance of communicating with Congress and others in Washington to assist him when necessary, and he also provided numerous examples of getting out and talking to his Marines by travelling around the world to visit every Marine base and station. The other CMC specifically discussed open communication and emphasized talking to people face-to-face. They also highlighted the importance of listening and receiving feedback.

General Gray suggested that you must talk to people and ask them how they are doing, and to show them that you care about them. "I used to go around," he recalled, "and talk to a lot of Marines. You've got to talk to them." General Gray also talked to their families, which had a significant impact on his command climate and its influence on the ethical behavior of his Marines: "You've got to take care of people, and that's part of your climate, and talking to the wives and the kids when their units are overseas, telling them how things are going."

Similarly, General Conway emphasized the importance of taking care of the families and talking to the wives: "Hey, it's not just about the Marines. It's about the families, too." He suggested that a good command climate must include talking to the families and taking care of all the wounded warriors, which will also have a positive influence on the ethical behavior of the Marines in the organization. He also mentioned that commanders must talk to their Marines face-to-face: "Break it down, bring it in, sit, kneel, bend, okay? And then you've got eyeball to eyeball contact, and every one of them can see you and again, gain their own impressions of you as you're gaining your impressions on what you're saying, and how it's going over."

Likewise, General Krulak suggested that there are other people besides the troops who the commander must talk to. For example, the

Commandant must establish open communication with Senators and Congressman as part of the Marine Corps' command climate. General Krulak used the example of when the Marine Corps was directed by Congress to cut its end-strength from 200,000 Marines down to 159,000 Marines. General Krulak was serving under Commandant Carl Mundy Jr. at the time and asked him for permission to go to Capitol Hill and talk to Senator Dan Inouye about the "unassailable sale" (which was the sale to get the Marine Corps fixed at 174,000 Marines). General Krulak had spent 18 years in Hawaii as a child and knew the senator. So, when Krulak presented the issue to Inouye, a member of the Senate Appropriations Committee, Inouye realized that the drawdown would require the Marines to leave Hawaii, which would cost the Hawaiian economy millions of dollars. General Krulak explained that "the Senate and the House Appropriations Committees are the people who give you the money. And they ended up so impressed they gave us 177,000 Marines, which is where we were before going to *Desert Storm*."

Like General Krulak, General Hagee discussed the need for commanders to be able to communicate both internally and externally to their organization: "Commanders today, in my opinion, need to be able to talk. And they need to be able to talk both to their unit, and they need to be able to talk outside the Marine Corps." He also emphasized face-to-face communication when he used the example from his time in Vietnam, where he had a Marine from Alabama, who posted a Confederate flag outside of his tent, and one of the black Marines wanted the flag removed. General Hagee had the two Marines discuss the issue face-to-face until the Marine from Alabama removed the flag once he understood why it offended the black Marine: "The Marines provided the solution for the betterment of that particular command, and therefore, the command climate."

Both General Neller and General Amos expressed their concerns with open communication relative to the use of social media. General Neller expressed suspicion of social media: "People have become so wedded and dependent upon their phones to communicate that the ability to have conversation is something that doesn't happen." He published specific guidance to the Marines that covered the rules and regulations associated with the appropriate use of social media.[314] During the interview he discussed how the Marines have transitioned from many online and computer-based programs of instruction to classes covered by small unit leaders. "We've directed more training be done by NCOs with their units where they have personal relationships with the Marines that are there," he said, "So there's conversation." General Amos expressed similar concerns: "I was the first Commandant that really had to face the reality of this thing called social media. I don't think the message changes (i.e., values). I think maybe the methodology of making the communication must be addressed. And for me, it was face-to-face."

General Dunford emphasized the importance of open communication when he said, "Now is not the time for the strong silent type, right? I don't actually think we should have the strong silent type in command. You've got to be the communicator-in-chief." He suggested that commanders must be able to articulate their expectations of behavior to their Marines and that the commanding officer's personal example must be accompanied by talking to his or her Marines: "It's the way that the commander communicates the standards that are expected inside of that organization." Correspondingly, General Amos commented that open communication is instrumental to a successful command climate.

For example, when he was asked to describe what a command climate looks like that encourages ethical behavior, he responded, "Number one, there had to be a constant line of open communication, absolutely, and it starts with, it starts with the commanding officer. It's got to be leadership communicating down and followers communicating up." General Amos brought up the time he visited Marines in Iraq and was escorted by a very junior NCO. The young Marine was a superb communicator and extremely confident in his abilities to converse with senior leaders. According to General Amos, this was a positive reflection on the climate

314. Neller, B. "Recent Conduct on Social Media." White Letter No. 1-17. Headquarters Marine Corps, Washington, D.C, 2017b.

of that organization and an example of open communication. In addition to talking to people, the CMC discussed the importance of listening and soliciting feedback as part of open communication.

Generals Gray and Conway both discussed the significance of critiques following military exercises, and that it is important to listen to your Marines to receive good feedback. General Gray mentioned that officers standing in front of Marines after exercises, lecturing them on what went well and what did not go well, does not provide the commanding officer with the best feedback: "There are times when you want to leave your rank behind, particularly at critiques and after exercises." General Gray wanted his Marines to have a chance to talk and that he would listen to them: "I encouraged people to talk, sergeants, corporals. Because I don't care whether you went to the left or to the right. I care what you thought about it." Similarly, Generals Conway and Jones emphasized listening and feedback with General Conway's concept of collective leadership and General Jones' belief on the importance of consensus building. "Early on," General Conway recalled, "I realized the value of what I think is formally termed democratic or collective leadership style. I've just tried to employ it throughout my entire career." He believed that other people's opinions counted, and he wanted to hear what they had to say. He gave an example of when he was the senior evaluator for an exercise where the unit being evaluated was led by Col John Ripley, USMC (Ret). Ripley was an icon in the Marine Corps, and a hero from the Vietnam War that earned him the nation's second highest combat decoration, the Navy Cross.

During the evaluation of Ripley's unit, everyone was saying great things about Ripley's organization, but General Conway felt obligated to offer his critique of the things that did not go well for Ripley's Marines. His critique was well received and indicative of the command climate that Ripley had established. Furthermore, he believed in listening to others for their good ideas: "I've always tried to make it better, and I think, again, that this whole idea of collective leadership—not just in commentary but in execution and demonstrating to people that you really mean it and you really do want to hear what they have to say, and you really do want to listen—is important."

General Jones said that "if you want something to survive your tenure, which is very brief, you've got to do a lot of consensus building, and you've got to make sure people agree that what you're asking them to do is actually worth doing." Like General Conway, he listened to others and wanted their opinions on how to make things better. He suggested that consensus building is one method to generate feedback and a way to get buy-in from your subordinates: "You have to provide the command climate that generates those discussions, not from the top down, but really from the bottom up." General Jones believes that one of the best compliments a leader can receive is when the subordinates' feedback confirms what the leader had planned or what the leader had thought about six months ago, is working.

General Hagee expressed strong feelings about the significance of listening: "You have to get the staff to where they will open up with you and have a conversation with you. And that means you have to keep your damn mouth shut and listen, and not shoot anyone in the face." General Hagee emphasized the importance of having people on your staff who will be very open and honest in providing the commander with advice: "You need to have someone like that. Because, if you don't, the perks, the benefits, guys want to please you. Just like being a battery commander, and telling the battery gunnery sergeant, I would like, and then, the next morning it appears."

General Dunford suggested that commanders who let their egos get in the way of leading their troops will impede their ability to listen to people and, deter Marines from offering good feedback. "You look at Chesty Puller," he offered as an example, "and what was the one thing that characterized him in all of the exchanges with young enlisted Marines? He was asking about their families. He was asking about their personal welfare." Even though Puller, a five-time recipient of the Navy Cross, was a larger-than-life figure, he was very humble when he talked to his junior Marines. According to General Dunford, humility is a necessary personal quality in attempting to establish the right climate to influence the ethical behavior of the Marines.

Another area that every Commandant emphasized during the interviews was core values and the importance of maintaining the intangibles that make United States Marines special.

3) Core Values

The CMC made various points regarding Marine Corps values and the Corps' high standards. The primary areas covered were the Marine Corps' high, almost spiritual, standards, the process of becoming a Marine, or what it means to become a Marine (The Transformation), and the significance of living up to those high standards. Although the CMC discussed the above topics, no Commandant emphasized the Marine Corps' high standards or its core values more than General Krulak: "The two most important things that the Marine Corps does for the nation are to make Marines and to win battles." He focused especially on the former and said that it is the individual Marine's values that are key to the Corps' success and critical to its survival.

General Krulak assisted commanders with building the foundations of their command climates by publishing a series of messages to all Marines, which covered subjects linked to the Marine Corps' core values and the Corps' high standards such as integrity, courage, and fidelity. These messages were designed to supplement General Krulak's *31st Commandant's Planning Guidance*, in which he discussed core values: "Our Marines' moral character, courage, and ethical values will dominate any location or operational area with unconditional certainty that the Marine Corps is a force for good."[315]

He suggested that Marines should be flexible, adaptive, innovative, and able to fight across the full spectrum of conflict as moral warriors. As he put it, "Marines must be seen, respected, and depended upon." In other words, the Marine Corps' core values are not to be interpreted as just a

315. Krulak, *31st Commandant's Planning Guidance*, 1995.

saying. General Krulak expected "them to frame the way we live and act as Marines." Similarly, both Generals Neller and Amos discussed the Corps' high standards and the importance of the core values.

"I think if we are that Marine Corps that's willing to appeal because we are different," he said, "and we hold people to a higher standard, that we expect a higher level of performance, that we expect a higher level of integrity, of honor, courage, and commitment. Then the Marines will continue to be able to recruit good citizens who will make good Marines." Likewise, General Amos suggested that the high standards Marines hold is a "spiritual thing": "That is the stuff that is the visceral deep-down-in things that cause us to be absolutely fearless on the battlefield, and willing to give our lives to the Marines to the left or the right."

When General Krulak discussed the rationale behind developing the Crucible, he explained, "the Crucible was to set the stage for what it is to really be a United States Marine, to live up to the high, almost spiritual standards of being a Marine." As he put it, "The mother and father are not going to sign on the dotted line to send their kids to the Marine Corps if we don't meet their almost spiritual standards."

Interestingly, General Gray correlated moral behavior and values with command climate and various methods that can be employed by commanders to get their moral message across to the Marines: "It's kind of the art of getting things done as opposed to the science of getting things done. And the art part of it comes from little by little, without talking a lot about it, without hardly saying anything about it, but you let the people figure out what your moral philosophy is about things and about what's right and what's wrong."

General Conway emphasized the importance of core values and for Marines to always do the right thing, especially when you are the leader. Once he had to relieve a general officer for misconduct, but the Secretary of the Navy did not want Conway to go through with the demotion of the officer, because the Secretary knew the officer and liked him. General Conway recalled, "So I got a great deal of pushback from the Secretary on trying to do the right thing with the officer in question. It took me three trips. But he just almost refused to allow me to do the right thing." General Conway called it "the Lance Corporal Rule," as in, "if a lance corporal gets punished for doing this, a general gets punished for doing the same thing." Conway upheld the Corps' high standards by relieving the senior officer, who was forced to retire. Three other CMC discussed the Corps' high standards.

General Jones said, "It's not what people do when people are watching. It's what you do when people aren't watching." General Hagee discussed his observations of other commanders throughout his career: "The guys I've seen fired as commanders were really outstanding officers. And most of them had an ethical lapse. They started to believe some of the stuff that they were above it, and that ended their careers." General Dunford suggested that one of the most important things our commanders can do is to communicate the Marine Corps' high standards to their Marines when he commented that "it's the way that the commander communicates the standards that are expected inside the organization." In addition to the high standards of the Marine Corps, which include the core values of honor, courage, and commitment; commanders are also responsible for making their Marines better people, which is a key aspect of the Transformation and the command climate.

General Gray mentioned the message from General John A. Lejeune (13th Commandant) which suggested that it is the responsibility of the commanding officer to make their Marines better people: "And that ought to be the guidance from the word go. And better people make better citizens in terms of the good of the country." Likewise, General Krulak discussed the need for the Marine Corps to possess good people based on the chaotic nature of the global environment and the places where Marines will be sent to conduct military operations. "We're in a world of chaos," he noted, "And if you're going to be in a world of chaos, you'd better be able to tie that individual Marine to something more than just the ability to pull a trigger and hit a target, because they're going to be making decisions that are going to change the tide of how the battle is fought. They're going to be the strategic corporal."

He used the example of the tragic events that unfolded with the U.S. Army at Abu Ghraib prison in Iraq, and the young female corporal who was photographed next to the hooded Iraqi prisoner. This was a shameful experience for the U.S. military. Instead of receiving the support of the surrounding Arab countries like the military had received during *Desert Storm*, the Arab countries turned against the U.S., which ultimately changed the landscape of the strategic picture in the Middle East. This was also a contributing factor in the birth of the insurgency in Iraq with organizations such as ISIS.

Both Generals Krulak and Amos emphasized the importance of character and how we recruit, train, and develop Marines, which became known as the "Transformation." General Krulak said that "because Marines are the centerpiece of the Corps, how we recruit them, train them, instill in them our core values and a sense of integrity and accountability, equip them to do the job, and treat them with dignity, care, and concern must be a principal emphasis."

He developed the idea of the Transformation, which accounted for the chaotic nature of the asymmetric battlefield, the Crucible through which civilians are transformed into Marines, the idea of the strategic corporal, the concept of the three-block war, and the needs of the Generation X recruits coming into the Marine Corps, who wanted to be part of something with values that were bigger than themselves.

All of these require a Marine Corps that possesses people with values and have superior character. As General Krulak put it, "What we want is ethical Marines. We want them to be thinking Marines. We want them to do the right thing, at the right time, for the right reason, when nobody's watching. And at the end of four years or 40 years, we're going to send you back to the same place you came from, better for having been a Marine." This was General Krulak's promise to the parents of a Marine recruit and his promise to the nation.

General Amos focused his remarks on the importance of core values and getting back to our roots as Marines when he delivered the Heritage and Values brief to Marines in 2012: "The Marine Corps can't allow itself to become like everyone else" (Amos, 2012c). In his *35th Commandant's Planning Guidance* he also emphasized that Marines are guided by the Marine Corps' core values: "These core values have been the compass for every Marine's service throughout our rich history."[316]

General Amos suggested that Marines must get back to good order and discipline. "We are allowing our standards to erode," he cautioned.[317] He was the only Commandant to put out a white letter on command climate where he emphasized that it is the responsibility of the commanding officer to ensure Marines are abiding by the core values and doing what's right: "A commanding officer has the greatest influence in determining whether the Marines in the unit are combat ready, whether they're honorable, whether they're forthright, whether they're focused, and how/whether they uphold our values and virtues."[318] Both Generals Krulak and Amos suggested that the NCOs play a significant role in sustaining the Transformation.

General Krulak used the term "Power Down," which emphasized small unit leadership as being key to sustaining the Transformation and the character development of the individual Marine. He advised[319] that "Everybody is going to have to spend some time during the week talking leadership. And we're going to make sure that it goes down to the very lowest denominator. So, we had something called Power Down." Like General Krulak, General Amos focused on command climate and the support from NCOs to assist the commander in his duties. He launched the "Reawakening" campaign, which was designed to remind NCOs of their critical role in making sure Marines live up to the title of Marine: "Fully 83.06% of our enlisted force is led by NCOs. And I need every one of you in this fight. Never forget *who we are* and *what we do* for our country."[320] In addition to the NCOs, General Amos delivered specific

316. Amos, *35th Commandant's Planning Guidance*, 2010, p. 1.
317. Amos, J. "Leadership and Conduct." White Letter No. 1-12. Headquarters Marine Corps, Washington, D.C., 2012a, p.1.
318. Amos, J. "Command Climate." White Letter No. 3-13. Headquarters Marine Corps, Washington, D.C., 2013a, p.1.
319. Krulak, C. "Sustaining the Transformation." White Letter No. 3-98. (1998).
320. Amos, J. The Reawakening. Headquarters Marine Corps, Washington, D.C., 2013b.

guidance to the commanding officers on the significance of conduct and the importance of the Marine Corps' high standards. "We lead by example," he said, "and provide continual and close supervision to those we have the privilege to lead. I expect each of you to hold yourselves and your Marines to the highest standards. Nothing else is acceptable."[321] He suggested that commanders have various tools available to assist them in maintaining the high standards in the Corps by using the Marine Corps Martial Arts Program (MCMAP) and junior leadership to reinforce discipline, values, and to promote character development.

General Jones recommended the MCMAP as a way for Marines to maintain good order and discipline and to ensure that Marines live by the core values. He relied on his experiences from Vietnam and his memories of how the Korean Marines kept their discipline and integrity. He mentioned that he realized good results from martial arts training by using the NCOs and junior officers: "We let the NCOs and the young officers develop it and make it into something that was usable in the Marine Corps. But almost immediately, the ethical behavior of the battalion changed." The Marines looked upon the belts associated with the MCMAP as a badge of honor and understood that they could lose their belt for incidents of misconduct.

Generals Hagee, Dunford, and Neller all suggested that commanders utilize the Socratic method as part of their command climate to ensure Marines stay in touch with the ethical issues required to sustain the Transformation. "By having that discussion," explained General Hagee, "you set up better individuals that make better decisions, which is going to help the command climate."

General Dunford emphasized the importance of command climate and sustaining the core values and maintaining the high standards. "Once you get beyond recruit training," he pointed out, "it is all about the commander establishing the proper climate within which our core values are maintained." He suggested the use of the Socratic method

321. Amos, J. "Leadership and Conduct." White Letter No. 1-12. Headquarters Marine Corps, Washington, D.C., 2012a.

to get Marines to think through problems: "Socratic in the sense that you're saying, 'Hey look, here are the facts. Talk to me about your decision making under these circumstances.'"

General Neller advocated for values training to be conducted by the NCOs: "I've directed this training be done by NCOs with their units where they have personal relationships with the Marines. So, there's conversation."

The Marine Corps' core values and its high standards enable Marines to maintain the trust of the American people. During his interview, General Krulak expressed the conviction that America believes Marines are good for the country. He quoted his late father, Victor Krulak, who said, "Should the American people ever lose that conviction as a result of our failure to meet their high, almost spiritual, standards, the Marine Corps will then quickly disappear."[322]

During their interviews, the CMC acknowledged that America does not need a Marine Corps, but America wants a Marine Corps. Therefore, to keep the trust of the American people and to ensure the survival of the Corps, it is incumbent upon all Marines to continue to uphold the high standards of the American people and the high standards of the United States Marine Corps. As General Gray said, "The nation expects you to be special. That's the reason that you have a Marine Corps because the nation wants you to have a Marine Corps, through Congress."

When General Krulak mentioned the need for a Marine Corps, he declared that "the American people want one." He also suggested that the conduct of the individual Marine is just as important as winning America's battles: "It's not whether you can win the battles that counts. It's do you win the hearts and souls of the American people." General Krulak shared the example of when two Marines and a Sailor in 1997 raped a twelve-year-old Okinawan girl. General Krulak flew across the

322. Krulak, V. *First to Fight: An Inside View of the U.S. Marine Corps.* Annapolis, MD: Naval Institute Press, 1984, p. xv.

world to talk to all the Marines on Okinawa about the incident and to reinforce the Marine Corps' values and standards.

Before leaving the island and after speaking with the Marines, he received word that Governor Ota wanted to meet with him. As General Krulak approached the Imperial Palace, he noticed it was lined with hundreds of photographers waiting to see what would happen next. When General Krulak, a Christian, walked up to the Governor of Okinawa, the governor reached out to shake his hand. General Krulak grabbed the governor and hugged him. General Krulak suggested that this act enabled the Japanese to save face and, as he put it, "Really defused the people of Okinawa. But a simple act like that can have an amazing impact because they didn't throw us off Okinawa." General Jones, for his part, addressed the need for Marines to uphold the special trust that America gives its Marines: "You should never do anything that embarrasses yourself, your family, your unit, or your country."

Generals Hagee and Dunford emphasized that the Marine Corps values life and that we must take our American values with us wherever we go. General Hagee recommended that commanders have difficult ethical decision making discussions with their Marines as part of their command climate, but discussions that are real and relevant issues that affect Marines. "That is part of helping to set," he explained, "the command climate to where individuals can think through that." He used the example of the dying wife and the poor husband who had an opportunity to steal a very expensive drug that would save her life. The ethical dilemma is the value of life juxtaposed against the husbands' values and that stealing the medicine is wrong. It is an ethical dilemma, but the right choice is to steal the drug because the value of life takes priority over the moral issue of stealing, even though many people make the mistake of not stealing the drug and allowing the wife to die.

General Dunford declared that the way America will eventually defeat organizations such as ISIS is that American military forces must bring American values to the fight: "Only by bringing our values with us and representing what's best in the United States of America are we ever going to be able to compete with the perverse ideas that we're dealing with. The issue for us in the long term is can the United States after all these years of war maintain the ethical standards that define us." General Conway discussed the importance of taking care of the wounded warriors and the families by establishing the Wounded Warrior Regiment, which also contributes to winning the hearts and minds of the American public: "We need to make these people feel like once they're wounded, they're not set aside—they just joined another unit."

Generals Amos and Neller both discussed the importance of upholding the Marine Corps' core values and keeping the trust of the American people. General Amos suggested that the soul of the Marine Corps is linked directly to the core values: "And that's what really gets to this thing called the soul of the Marines, the soul of the Corps. And that's the thing we can never, ever, ever allow to change." Also, he discussed the importance of upholding the standards relative to the survival of the Marine Corps. "The high regard of our fellow citizens," he cautioned, "and our own self-image are at stake."[323]

General Neller expressed his frustration with recent incidents of misconduct in the Marine Corps, which stemmed from the abuse of alcohol. In his interview, he recommended for all Marines to abide by the Corps' high standards and values: "Read more, drink less, PT smarter." He suggested that Marines are reading more and that they are making better decisions on their physical fitness. However, he also lamented that some Marines are not drinking less, which has caused many of the misconduct problems in the Marine Corps that make the news and erode the trust of the American people.

General Neller emphasized the importance of a Marines' oath to support and defend the Constitution of the United States: "This is not supposed to be easy. This is hard. You took an oath. It doesn't have an expiration

323. Amos, J. "Leadership and Conduct." White Letter No. 1-12. Headquarters Marine Corps, Washington, D.C., 2012a, p. 2.

date on it." Every Marine has an obligation to fulfill his or her oath to defend the Constitution and, as such, abide by the rules and regulations that govern the conduct of a Marine along with the values that are imbued in every Marine. According to the CMC, when Marines do these things, they promote greater trust with the American public.

4) Accountability and Responsibility

Another common category that originated from the interviews with the CMC was accountability and responsibility. Each Commandant discussed the significance of commanders holding themselves accountable first, and that they must hold their Marines accountable for violations of Marine Corps standards and the core values. The CMC also argued that it is the responsibility of the Commandant, the general officers, and all the commanding officers to protect the Marine Corps as an institution.

General Gray discussed how leaders must set the example and hold themselves to an even higher level of accountability, which requires self-discipline, "The nation expects me to be special," he said, "And if I can't live up to that, I can't be a Marine. And so I think it starts with self-discipline, and I think the Commandant's real solution is leadership." General Hagee suggested that one of the best ways for commanders to ensure they are holding themselves accountable is to continually evaluate their leadership and command climate: "What I think the better leaders do, is you're constantly thinking, evaluating your leadership, learning from both good and bad examples." Leaders must exercise the self-discipline necessary to constantly reflect on their own leadership and the type of environment they set within their organization. General Hagee added that leaders must look themselves in the mirror and ask the question, "Is that the person I want leading me?"

General Jones emphasized the concept of self-discipline when he suggested that a commander's climate is not working when "there is reckless behavior, evidence of a lack of discipline." He used the example of the MCMAP to act as a "badge" of discipline, to promote greater self-discipline and reduce incidents of misconduct for all ranks.

General Conway mentioned the importance of discipline when establishing the right command climate when he declared "discipline is important in a unit." He suggested that if the commander wants to set the right environment, it must be of a nature where Marines exercise self-discipline and that they understand there will be rewards for good behavior and punishment for bad behavior. General Amos underscored the significance of commanders establishing a climate where the organization has self-discipline, takes care of people, and does things the right way: "I hold them accountable for it."

Likewise, General Neller discussed how it is the commander's responsibility to hold him or herself to an even higher level of accountability when leading others: "Other than to say that the commander's got to role model the behavior that he expects every other Marine to exhibit. And he's got to hold himself more accountable than anybody else in the unit." General Neller mentioned that the followers have certain expectations of their commanders: "They expect us to hold people accountable, ourselves first and foremost."

When General Krulak discussed accountability, authority, and responsibility, he warned that "the one thing that they can't do is delegate the accountability." The commanding officer is ultimately held accountable for whatever his or her unit does or fails to do. General Dunford expressed a similar point, that "when you're a commander you are not responsible for the individual criminal conduct of any one individual. You're responsible for the climate within which those individuals operate every day."

General Dunford argued that when there is an incident, it is more important how the command or the institution responds to the incident than what happened. He also emphasized that it is the commander's command climate that sets the tone for the organization and that the commander must hold individuals accountable for their actions.

General Amos suggested that "if commanders set the right climate then many of the other issues take care of themselves, when you've got the right command climate, the right standards, the right things that have been told to your youngsters and the expectations, and you hold them accountable to it. I would say if there are a hundred bad things that can go wrong, 95 of them will be taken care of by good decision making."

Generals Krulak and Conway both used examples of holding senior leaders accountable. General Krulak remarked that senior leaders do not plan to make mistakes. Mistakes occur, in his view, when leaders let their moral compass waiver or ethical behavior is no longer their priority, and they fail to hold people accountable: "They're dead. They've lost the moral authority to take action. The most important authority you have is not the four stars on your collar. It's the moral authority you have in your soul." He also brought up the importance of holding general officers accountable. "There's no room in the Marine Corps for situational ethics or situational morality," he declared. "Those found wanting will be held accountable. I threw out two two-star generals, for not doing that. They were screwing their secretaries."

Similarly, General Conway had to relieve one of his senior leaders for misconduct: "We used to call it the 'Lance Corporal Rule,' any time I had to deal with a general. If a lance corporal gets punished for doing this, then a general gets punished for doing the same thing." He emphasized that we must hold all Marines accountable for violations of our core values and standards regardless of their rank or position.

General Hagee was explaining accountability when he recalled the impression of the followers when leaders hold people accountable: "Holy cow, he's really serious. If I cross that ethical line, then he is going to hold me accountable, and I don't want to do that." He used as an example the case of a Marine under his command who was caught for shooting some horses, and everyone was waiting to see how General

Hagee was going to handle the matter: "I court-martialed him. And that signal went out to everyone in the division. Uh-oh, we'd better do what's right." If he had not held the Marine accountable, he would have set a new standard in his division that would have made it difficult for him to punish Marines in the future for other violations of the high standards.

General Dunford expressed a similar sentiment: "But at the end of the day, I think if you look at how the institution responded in holding people accountable for aberrations of our core values, that to me is a reflection of the institution and the expectations of the institution." He also suggested that a commander can use their climate as a tool to reinforce the message that all Marines receive in boot camp and at officer training:

> Keep our honor clean. It is not keep my honor clean, right? It's not keep your honor clean. It's keep our honor clean. So, it's the sense of individual accountability to the institution and the sense of individual actions actually reflecting the institution's core values. So, from the very earliest days in recruit training, you are instilling in a private the sense of accountability to the values of the institution and the sense that if you violate those values, it's not about you. You are actually doing something that is putting a black mark on the institution as a whole, and that's particularly true the more senior you become.

General Dunford would tell his new one-star general officers that they had to be very careful as to what they said and how they behaved in public: "You can never again speak in public without being a reflection of the institution. You don't have a personal opinion in public anymore. And when you are communicating publicly, you actually don't have a personal opinion because you are the institution."

During his Reawakening Campaign, General Amos directed the NCOs in the Marine Corps to assist him along with their commanders to combat the rash of misconduct events in 2013: "Move to the decisive point in this battle and through your presence, professionalism and tenacity…turn the tide of this fight for the sake of Corps and country."[324] For General Amos, the very existence of the Marine Corps as an institution was in jeopardy.

During his campaign against sexual assault and hazing, he discussed leadership and his responsibility to protect the Marine Corps "As Commandant," he said, "I have no greater responsibility to our institution than to ensure that our Corps adheres to the legendary high levels of discipline and professionalism expected of us by the American people."[325] In his Heritage and Values brief to Marines, he quoted then-Commandant Carl Mundy, Jr., 30th Commandant of the Marine Corps, who said to General Amos "you can't lose the spiritual health of the Corps as the 35th Commandant."[326] It is the responsibility of the Commandant to ensure that the Marine Corps remains relevant and that the Marines continue to be wanted by the American people.

General Gray would say to Marines "the nation expects you to be special. That's the reason that you have a Marine Corps is because the nation wants you to have a Marine Corps." Likewise, General Hagee suggested that it is the responsibility of the commander to protect the institution when he used Admiral Nimitz as the epitome of taking care of the needs of the Navy before his own personal goals.

Nimitz refused to take the position of Chief of Naval Operations over other more senior admirals at the behest of President Roosevelt. General Hagee suggested that one of the best leadership qualities of Admiral

324. Amos, J. The Reawakening. Headquarters Marine Corps, Washington, D.C., 2013b.
325. Amos, J. "Leadership." White Letter No. 3-12. Headquarters Marine Corps, Washington, D.C., 2012b.
326. Amos, J. "Heritage and Values Brief [Power Point Slides]." 2012c.

Nimitz, and one that was indicative of the climate he set within his organization was selflessness: "It was never about him." Nimitz was always protecting the Navy as an institution and he was always looking out for his Sailors before taking care of himself.

General Jones discussed the environment and how commanders must always be aware of their environment. The CNN effect is real and can create false perceptions about the Marine Corps. In his view, "the first thing that you have to do in our training of officers and staff NCOs is to make sure they understand that they are not invisible. As a matter of fact, they are visible 24/7, 365." As Generals Dunford and Hagee also argued, Marines in leadership positions must be particularly sensitive to their words and actions, which may reflect upon the Marine Corps as an institution.

General Krulak emphasized the connection between the climate the Commandant sets for the Marine Corps and his responsibility to the Corps as an institution.

"What I learned early on," he recalled "was just what my father wrote, and that is that your responsibility (as Commandant) is to the American people. It's do you win the hearts and souls of the United States, the American people." He suggested that to continue to win the hearts and souls of the American people, Marines must continue to uphold their legacy: "Our Marines' moral character, courage, and ethical values will dominate any location or operational area with the unconditional certainty that the Marine Corps is a force for good."[327]

In addition, General Krulak explained how the nation's trust is the Marine Corps' inheritance, which must be sustained as a sacred responsibility: "It is a debt we owe to those who have gone before us, and a promise we make to those who will follow. It is the guiding light of our

327. Krulak, *31st Commandant's Planning Guidance*, 1995.

ethos."[328] The nation demands that its Marines be always ready, capable, and that Marines possess the values of honor, courage, and commitment. General Krulak quoted the former 18th Commandant, General Alexander Vandegrift, who said "the nation has placed a measure of its trust and hope in the one hundred thousand men who have volunteered to serve the cause of freedom as United States Marines. The Marine Corps is always ready to fulfill that trust."[329] The CMC, Marine general officers, commanding officers, and all Marines are responsible for the stewardship of this sacred trust, which is carried out through living up to the reputation established by the Marines who have gone before, always doing the right thing as a force for good.

General Krulak quoted the 13th Commandant of the Marine Corps, General John A. Lejeune, when he emphasized the significance of the command climate established by all commanders and passed on to the next commander, to include the Commandant: "This high name of distinction and soldierly repute, we who are Marines today have received from those who preceded us in the Corps. With that, we also received from them the eternal spirit which has animated our Corps from generation to generation and has been the distinguishing mark of the Marines in every age."

General Krulak discussed the Transformation as an event that changes a person forever and instills in them the Marine Corps' beliefs, ideals, and values of honor, courage, and commitment. Krulak believes that the key to the Marine Corps' survival is for the individual Marine to always live by these core values, which will help to maintain the trust of the American people. To reinforce his point, Krulak quoted President John F. Kennedy:

> And when at some future date the high court of history sits in judgment on each of us, recording whether in our brief span of service we fulfilled our responsibilities to the state, our success or failure in whatever office we hold, will be measured by the

328. *Ibid*, p. 1.
329. *Ibid*, p. 1.

> answers to four questions: First, were we truly men of courage. Second, were we truly men of judgment. Third, were we truly men of integrity. Finally, were we truly men of dedication? [330]

Cross-Case Comparison Key Themes

There were three key themes that emerged from the cross-case comparison. The first is that the command climate set by the commanding officer is significant with respect to influencing the ethical behavior of the Marines in the organization. The second is that the commanding officer's climate must be focused on the Marine Corps' core values, and the character development of the individual Marine post entry-level training (sustaining the Transformation). Finally, the third is that the best way for the United States Marine Corps to continue winning the hearts and souls of the American people is through the ethical behavior of the individual Marine, both on and off the battlefield. This is crucial to the Marine Corps' survival as an institution. The three themes are discussed in greater detail below.

Theme One: Command Climate and Ethical Behavior

> Theme One: The command climate set by the commanding officer is significant with respect to influencing the ethical behavior of the Marines in the organization.

The first theme that emerged from the cross-case comparison of the CMC interviews is that the commander is solely responsible for setting the right climate in his or her organization. This requires commanders to set the example in everything they do through their own personal actions, that they hold themselves accountable, and that they hold their Marines accountable.

The climate of the organization emerges from both the commanding officer's behavior and the expectations that the commanders set for their subordinates. The policies, practices, and intentions that stem

330. Krulak, C. "Character." *Leatherneck Magazine*, 79, no. 12 (1996a): 14-15, p. 3.

from the command philosophy all have an impact on the climate of the organization and the ethical behavior of the people in the organization.[331] It is the commander's responsibility to establish an environment that not only sets the individual Marine up for success but also sets up the unit and the institution for success.

The first theme suggests that the command climate established by the commanding officer is critical in influencing the ethical behavior of the Marines in the command. General Krulak discussed his experiences in Vietnam and his former battalion commander who established a superior command climate. "He espoused all the things that I'm trying to get across to you," expressed General Krulak, "the idea of the importance of what is command climate in your individual unit. What do they think? How do they act? And, how do they respond?" General Amos "published a paper to all my general officers, and actually, to my commanding officers as well, about the importance of command climate, the importance of having the right tone set by the commanding officer is critical to the behavior of the organization."

General Dunford suggested that it is the responsibility of the commanding officer to take care of the "intangible qualities" of the unit: "I think it (command climate) is the primary driver of ethical behavior." General Neller emphasized that the commander must lead by example: "I think climate is an outcome of leadership and people that we lead expect us to be competent, committed, fair, and to be somebody of virtue and character." General Jones agreed: "If you don't have a good command climate you don't do anything well, let alone ethical behavior. So, I think it is one of the most important missions the commander has to deal with up front." General Jones suggested that the ethical behavior of the individual Marine is a key element of the Transformation process of turning civilians into United States Marines and making them better citizens.

331. Erhart, M., Schneider, B., and Macey, W. *Organizational Climate and Culture: An Introduction to Theory, Research, and Practice*. New York, NY: Routledge, 2014.

Theme Two: Sustaining the Transformation

> Theme Two: The commanding officer's climate must be focused on the Marine Corps' core values, and the character development of the individual Marine post entry-level training (sustaining the Transformation).

The second theme produced from the cross-case comparison is that the commanding officer's climate must include a heavy emphasis on ethical behavior based on the Marine Corps' core values and the high standards that govern the conduct of United States Marines. As previously discussed, it is the responsibility of the commander to regularly reinforce the core values and the importance of one's character to ensure that their Marines are better people because of their service in the Marine Corps.

Values reinforcement training and education must be conducted at the lowest level possible within the command structure. Ethical dilemmas should be introduced using the Socratic method of questioning to develop solutions to ethical issues. The study revealed that current Marine Corps doctrine and ethics curriculum today is more focused on values-based training, which aligns well with the core values and is taught to all ranks.

The second theme advocates for the sustainment of the Transformation within Marine Corps organizations after the Marine has completed his or her entry-level training. It is the responsibility of the commanding officer to ensure his or her Marines are better people than before they came to the unit. General Gray said, "I think General Lejeune's message had a great impact on me in terms of my thinking as an officer. Particularly, Lejeune's message about you owe it to the young people to make them better morally, ethically, and physically, when they leave you than when you joined them."

General Dunford agreed: "Once you get beyond recruit training it is all about the commander establishing the proper climate within which our core values are maintained." During his interview, General Krulak also brought up the significance of commanders sustaining the Transformation:

> But the intent of the *Transformation* was that you would continue that education (values training), that you would sustain the Transformation through at least weekly discussions. Even if it's 15 minutes, just one little vignette and sitting around the table, and here's the vignette that's going to be x, y, or z. When you get to the majors or to the company commanders, or to the Expeditionary Warfare School, or the Command and Staff College, or the MCWAR, at that point that education must become far more pointed. I mean, I would start it with that quote (Victor Krulak's quote), and I would use that as a pile driver into every one of their hearts and every one of their souls. Because that is what the American people expect of us.

General Hagee affirmed that commanders must sustain the core values and reinforce character development through the use of ethical dilemmas: "But setting that command climate, and then checking what you're doing to make sure you're doing everything that you can to reinforce the climate that you're trying to set. And to understand ethical, what I would call ethical dilemmas. Where there is good on both sides, and there's bad on both sides."

General Amos emphasized the importance of sustaining the Transformation when he discussed the overall health of a Marine Corps unit, which is a reflection on the command climate and has a significant influence on the behavior of the individual Marine:

> These healthy units, these well-led units, these units with high morale, these units that have great command climates, absolutely have fewer discipline problems. Absolutely have higher reenlistments. Absolutely are happier units. The units that have a healthy command climate are happy units.

General Jones suggested that commanders must work on reinforcing ethical behavior and that it must be done on a routine basis: "You have to keep reminding people because new people are coming all the time. If they don't hear that then all of a sudden the first thing that happens is you will have a major training, a major hazing incident. You have to keep talking about it."

Theme Three: Winning the Hearts and Souls

> Theme Three: The best way for the United States Marine Corps to continue winning the hearts and souls of the American people is through the ethical behavior of the individual Marine both on and off the battlefield. Again, this is directly related to the Marine Corps' survival as an institution.

The Marine Corps is known for its ability to make Marines and fight and win the nation's battles. Clearly, it must continue to accomplish these tasks to remain relevant as a warfighting institution. The results of the cross-case comparison suggest that the individual character of the Marine and the values the Marine Corps instills in its Marines contribute to warfighting effectiveness. In addition to the Marine Corps' combat record, it should also be known as the world's premier character development institution.

The third theme underscores the significance of the commander's climate and the ethical behavior of the Marines by winning the hearts and souls of the American people. This theme suggests that Marines must live up to the expectations, the legacy, and the conduct associated with being a United States Marine. These are the lofty (almost spiritual) expectations of the American people. These themes are directly connected to the longevity of the Marine Corps, since America already has a very capable land force in the U.S. Army and a very capable air force in the U.S. Air Force. America does not necessarily need a Marine Corps, but America wants a Marine Corps. The CMC suggest that the reason Marines continue to be wanted is because of their behavior and living up to their legacy as a force that is good for the country.

General Jones discussed how the behavior of the individual Marine is critical in the eyes of the American public and is a reflection on the Marine Corps as an institution: "You should never do anything that embarrasses yourself, your family, your unit, or your country." General Neller underscored the importance of winning hearts and minds with an example from a recent incident in Okinawa: "We've got three Marines who did some silly things on Okinawa this past week. That's three out of 22,000. And then, because of those three, it's easy to say the climate sucks in Okinawa. No, it doesn't."

General Krulak rejoined that America does not need a Marine Corps, but rather "the American people want one." Why? General Krulak answered this question when he discussed what he called "the business of the business" of the Marine Corps. In addition to discussing the business of recruiting, recruit training, manpower, and sustaining Marine Corps values, he also reflected on the lineage of the Corps. He pointed to great leaders such as General Holland "Howlin Mad" M. Smith, whose vision of amphibious operations in the Pacific took form while he was serving as a young officer staring out at the wheat fields (i.e., the ocean) of Belleau Wood in France during World War I, while preparing to attack the "Pacific islands."

General Krulak also referred to the island fighting battles in the Pacific and other great Marine Corps leaders such as Vandegrift, Twinning, and, of course, his father, Victor Krulak. His point was that Marines have been doing this business of fighting and protecting America for over 242 years, and that the Marine Corps has never let down the country which it proudly serves. Along with the Corps' illustrious history, General Krulak suggested that all commanders must thoroughly understand the 1947 National Security Act and its description of the mission of the Marine Corps and the speech (Bended Knee) given by then-Commandant Vandegrift. Vandegrift's speech addressed a Congress that was ready to disband the Marine Corps after World War II. These are all examples of the "business of the business" that Krulak referred to during the interview. General Krulak argued that the conduct of the individual

Marine is equally as important, or even more important than the Corps' ability to win every battle that it fights:

> I am telling you, we could have gotten the *shit* kicked out of us in Fallujah. And so long as we fought hard and we were honorable, we're fine. But take a picture of Marines urinating on enemy dead, that's not good, that's not good. Having Marines take pictures of female Marines nude or in a state of undress, and then posting them on Facebook with derogatory words, that's not good.

His father spoke about the significance of winning the hearts and souls of the American public:

> We exist today, we flourish today, not because of what we know we are, or what we know we can do, but because of what the grass roots of our country believe we are and believe we can do. The American people believe Marines are downright good for the country; that the Marines are masters of a form of unfailing alchemy which converts unoriented youth into proud, self-reliant, stable citizens—citizens into whose hands the nation's affairs may safely be entrusted. And likewise, should the people ever lose that conviction as a result of our failure to meet their high, almost spiritual standards, the Marine Corps will then quickly disappear.[332]

General Amos supported this view:

> I don't think anything has ever been written that's as poignant and as enduring as what he (Victor Krulak Sr.) described. If your behavior caused the American people to not believe in you, that you brought shame on your unit, and therefore, on the Marine Corps such that the American people no longer believed that

332. Krulak, V. *First to Fight: An Inside View of the U.S. Marine Corps.* Annapolis, MD: Naval Institute Press, 1984, p. xiv-xv.

> a Marine Corps was just downright good for it. I mean that's a measure that would self-correct.

General Dunford suggested that the people responsible for winning the hearts and minds of the American people are the leaders:

> It's keep our honor clean. It's you got to be able to trust your fellow Marines. Its officers eat last in the mess line. So, it's very practical ways that these things have - We are an oral tribal society, right? So, this is a couple hundred years of oral tradition. What John W. Thomason calls such things as 'Regiments hand down forever.' So, he is a World War I guy. He writes a book when he comes back home and talks about such things as, 'Regiments hand down forever.' What he was talking about was all these intangible qualities. And who did he say was responsible? The leaders. Passing those from one generation of Marines to another. So, that's why I think it is less about a program of instruction or a period of instruction then it is - it's part of the fiber of the institution.

Cross-Case Comparison Conclusion

After interviewing each of the CMC and then comparing the individual cases, the analysis suggests that the most important thing a commanding officer of Marines can do is to establish the right command climate. A command climate that is focused on the Marine Corps' core values of honor, courage, and commitment, the high standards that all Marines are expected to meet, and the individual character development of each Marine in the organization. Of course, the climate that the commander establishes will have a direct impact or influence on the behavior of his or her Marines and must be focused on the individual character of the Marine. For commanders to be able to influence the ethical behavior of their Marines they must primarily set the example through their own individual actions.

Commanders must also establish a climate that encourages the thoughts and ideas of others through open communication both up and down the chain of command. The commander's climate must be an environment where every Marine believes that he or she has a voice in the organization and that anyone can speak their mind if they do so in a professional, considerate, and respectful manner.

For the climate to have the most impact or influence on the ethical behavior of the Marines, the sustainment of the Transformation must be centered on the Marine Corps' core values and the ongoing character development of the individual Marine. This training and education cannot be relegated to traditional lectures from senior officers, but rather, driven by junior enlisted leaders such as NCOs and junior officers who are closest to the troops. These discussions should be weekly and must be based on the Socratic method where the leaders present an ethical dilemma to the troops and allow them to develop a solution without the leaders providing the answers.

Finally, for the Marine Corps to continue to win the hearts and souls of the American people, it must continue to not only perform as the world's premier warfighting force, but it must also be known as the world's premier character development institution, and that it is truly an organization that is good for the country. This behavior will continue to promote the trust of the American people that the Marine Corps has experienced for over 243 years.

CHAPTER FOUR

STUDY RECOMMENDATIONS

This chapter presents two recommendations that were generated from the three key themes—Command Climate and Ethical Behavior, Sustaining the Transformation, and Winning the Hearts and Souls:

1. a proposed Marine Corps definition for command climate and

2. a proposed command climate survey that would replace the command climate survey currently being used by the Marine Corps.

The qualitative procedures described in detail (see Appendix A) were helpful in addressing the research questions. The open-ended, semi-structured interviews provided rich data that was transformed into three themes that shape how each Commandant viewed command climate and the factors that influence ethical behavior within the command.[333]

Recommendations for a Definition and a Survey

The four common categories and the three key themes previously discussed were leveraged to construct a proposed definition of command climate for the United States Marine Corps. The proposed definition incorporates elements of the previously mentioned themes as a process climate focused on "internal processes that occur in organizations as part of daily organizational functioning such as, a procedural justice climate, or an ethical climate."[334] The definition emphasizes ethical behavior as the primary desired outcome of the command climate. In the proposed definition, the command climate puts a heavy emphasis on

333. Patton, M. *Qualitative Research, and Evaluation Methods.*
334. Erhart, M., Schneider, B., and Macey, W. *Organizational Climate and Culture: An Introduction to Theory, Research, and Practice*, p. 87.

core values and the regular reinforcement of ethical training and education to sustain the Transformation of the individual Marine.

The proposed definition includes the importance of winning the hearts and souls of the American public and that the command climate must prepare and groom Marines to live by, and demonstrate, the high expectations of the American public. The cross-case comparison suggests that the longevity of the Marine Corps will be determined by how well Marines live up to those lofty expectations. In the proposed definition of command climate, the commander sets the tone for the organization as well as the expectations for the subordinates. It is not about the actions of the commanders but rather the result of the actions of the commanders. Therefore, the environment does not simply "set Marines up for success, enabling them to accomplish their individual goals" but, equally important, allows the unit to accomplish its goals and enables the Marine Corps to accomplish its mission as an institution.

Proposed Definition of Command Climate

After analyzing the data generated in this study, the definition I propose is as follows:

Command Climate: The environment set by the commanding officer that enables success for the individual Marine, the unit, and the institution by promoting the Marine Corps' core values, enforcing its high, almost spiritual standards, and emphasizing the importance of living up to the historical expectations established by the American people who believe that Marines are good for the country.

The above definition charges leaders at every level with the responsibility for creating an environment that is grounded in the Marine Corps' values, ethos, and standards. This environment is where the policies, practices, and intentions of the leaders send a clear and consistent message to Marines about what is valued the most in the organization—a Marine's character. Further, this environment promotes setting the example, open communication, non-negotiable core values, and accountability and responsibility. True to the Marine Corps' maneuver warfare doctrine, this type of climate encourages people to treat one another with dignity and respect—generating trust among members of the organization. A climate focused on ethical behavior, values, and character ensures that commanders will hold accountable all violators of the Marine Corps' high standards and non-negotiable values. When these values are violated, they tend to erode the American publics' belief that "Marines are downright good for the country."[335]

Proposed Command Climate Survey

The second recommendation that emerged from the results of the study is the need for an update to the Marine Corps' Command Climate Survey.[336] A new survey updating the Marine Corps' existing command climate survey would provide new questions to gather respondents' perceptions of their units' command climate and 1) the role it plays on influencing their ethical behavior and 2) how it fosters character development. Participation results in this survey offers the commanding officer valuable feedback necessary for a commander to take the appropriate actions to improve the command climate and enhance warfighting effectiveness. The new survey will give the commander a snapshot of how well his or her command is accomplishing the goals and objectives of the organization and setting the right conditions for success—the individual Marine, the unit, and the institution. This survey incorporates questions based on the data from the interviews with the CMC while maintaining the original structure and integrity of the current survey.

This instrument uses a cross-sectional survey designed to capture a current snapshot of the member's perceptions of the command.[337] The cross-sectional survey design makes sense based on the routine, high volume of personnel turnover, and individuals assigned to temporary duty (e.g., attending professional schools and special assignments

335. Krulak, V. *First to Fight: An Inside View of the U.S. Marine Corps.* Annapolis, MD: Naval Institute Press, 1984, p. xv.
336. See page 151, this volume.
337. *Ibid.*

external to the command). The first group of questions consists of demographic information useful to commanders in analyzing their command climates and remains the same as the current survey produced by Headquarters Marine Corps. The next group of questions is associated with the unit's approach to character development, ethical decision making, and personal communication.

These questions are organized using a Likert scale[338] and are designed to gather feedback from respondents' perceptions of how well the command is doing reinforcing the Marine Corps' core values of honor, courage, and commitment through classroom instruction, group, and individual discussions (e.g., the Socratic method and ethical dilemmas), as well as practical applications designed to further develop the character of the men and women in the command. These questions stem from ethical climate theory and are also based on various other ethical theories such as values-based ethics, ethical fading, and a pedagogy based on the Socratic method.[339]

Following the nine Likert scale questions are six open-ended questions that enable the respondent to elaborate on their perceptions of the command's ethical training, leadership, and the degree of leadership's involvement with the individual Marine's character development. The open-ended questions request recommendations for making the unit a better organization. The survey concludes with three yes/no questions about the leadership of the organization and one question representing the frequency of ethical discussion training that provides the respondent with a sliding scale to depict their perceived degree of training frequency.[340] Finally, one question asks how well the unit sustains the Transformation.

The proposed survey is included on the following pages, for the Marine Corps to review and consider as a possible survey instrument for commanders. This tool will provide commanding officers with a "pulse check" on how well their command climate influences the ethical behavior of the Marines in the command, and how their command climate influences the character development of the members within their organization.

338. Fowler, F.J. *Survey Research Methods*. Thousand Oaks, CA: Sage Publications, 2014.
339. Socratic Method, 2018.
340. Fowler, F.J. *Survey Research Methods*.

COMMAND CLIMATE SURVEY

Adapted with permission from USMC.

PURPOSE

This survey is designed to give members of the organization the opportunity to provide anonymous feedback to the commander. The purpose of this survey is to gather respondents' perceptions of their unit's command climate and the role it plays on influencing their ethical behavior. Your participation in this survey provides the commanding officer with the feedback necessary for him/her to take the appropriate action to make improvements within the organization.

SURVEY INSTRUCTIONS

- All survey items must be answered.
- The survey should take 15-20 minutes to complete.
- Once all survey items are completed, click on the Submit button at the top of the page.
- When you have submitted the survey, a "Thank You" note will appear.

OUR COMMITMENT TO ANONYMITY

- Responses cannot be tied to the respondent. Your responses are anonymous. Your responses are separated from your personal information (e.g., rank, gender, etc.) so that your commanding officer cannot tie the two together. For example, you may be the only female, E-8 in your unit, but your commanding officer cannot link your responses to that information.

Q1. Please select one:

- ▫ Military
- ▫ Civilian

Q2. Please select your grade/rank:

- ▫ E1
- ▫ E2
- ▫ E3
- ▫ E4
- ▫ E5
- ▫ E6
- ▫ E7
- ▫ E8
- ▫ E9
- ▫ WO1
- ▫ CWO2
- ▫ CWO3
- ▫ CWO4
- ▫ CWO5
- ▫ O1
- ▫ O2
- ▫ O3
- ▫ O4
- ▫ O5
- ▫ O6
- ▫ O7-O10
- ▫ N/A

Q3. Please select your age:

- ▫ 17-20
- ▫ 21-24
- ▫ 25-28
- ▫ 29-32
- ▫ 33-36
- ▫ 37-40
- ▫ 41-44
- ▫ 45 or older

Q4. Please select your branch of service:

- ▫ USMC
- ▫ USN
- ▫ USA
- ▫ USAF
- ▫ Foreign Service
- ▫ USCG
- ▫ Prior Military Service

Q5. Please select your status and check all that apply:

- ▫ Active Duty
- ▫ Civilian
- ▫ Contractor
- ▫ Traditional Reservist / IMA
- ▫ Work Study, Intern
- ▫ Full-time technician
- ▫ Federal Civil Service
- ▫ Temporary employee

Q6. Please select your gender:

- ▫ Male
- ▫ Female
- ▫ Other

Q7. Please select your time in the unit:

- ▫ 1-12 months
- ▫ 13-24 months
- ▫ 24 months

Ethics is defined as the study between right and wrong, good and bad behavior. *Values* can be defined as "important and lasting beliefs or ideals shared by members of a culture about what is good or bad and desirable or undesirable." *Ethical climates* set conditions for appropriate ethical behavior, and positive outcomes are created from ethical practices such as integrity, trust, justice, and social responsibility. *Unit cohesion* is both vertical and horizontal. *Vertical cohesion* is the relationship between the seniors and the subordinates. *Horizontal cohesion* is the relationship between peers and personnel of similar rank (e.g., NCO's, SNCO's, etc.).

Q8. Please select an answer for each of the following statements:

	Strongly Disagree (1)	Disagree (2)	Agree (3)	Strongly Agree (4)	Don't Know (5)
1. Leaders/supervisors emphasize the importance of doing what's right.					
2. Leaders/supervisors participate in ethical discussions with the troops.					
3. Leaders/supervisors communicate down to the most junior level.					
4. I feel leaders/ supervisors are preparing me to make good ethical decisions.					
5. My unit has a character development program.					
6. The command climate sets a good example for ethical decision making.					
7. The command climate has a positive influence on my ethical behavior.					
8. The members of this unit trust the leadership/ supervisors.					
9. My unit has strong cohesion.					

Q9. Referring to the previous questions, if you have any additional comments, please provide them below in a sentence or two. Please reference the question(s) you are addressing:

____

____

____

Q10. What is the biggest area your unit could improve on to set a command climate that promotes ethical behavior?

____

____

____

Q11. Would you want to serve in this unit again? Why or why not?

____

____

____

Q12. What training could your unit provide that would prepare you to make good ethical decisions?

____

____

____

Q13. Describe how your leaders/supervisors provide mentoring on ethical behavior?

____

____

____

Q14. Consistent with the Marine Corps' core values, does this unit encourage you to be a better person? Why? or Why Not?

____

____

____

Q15. Do people in this unit correct each other when they notice someone is doing something wrong?

- ▫ Yes
- ▫ No

Q16. Does your commanding officer establish an environment that promotes mutual trust and respect?

- ▫ Yes
- ▫ No

Q17. Do the non-commissioned officers guide discussions on ethical decision making?

- ▫ Yes
- ▫ No

Q18. Ethical discussions occur in this unit:

	0	10	20	30	40	50	60	70	80	90	100
Frequently											
Occasionally											
Seldom											
Never											

The *Transformation* is the process that begins with the prospective Marine's first contact with the Marine recruiter and continues through the Marine's entire life. During the *Transformation*, Marines must continue to maintain a true compass of personal honor and commitment to forego interests of self for the interests of their comrades, their Corps, and their country. The legacy of the Marine Corps is ultimately judged by the entire *Transformation* experience of all Marines and that legacy influences the next generation of Americans that will join our ranks that will start the cycle over again. (For more information, see General Charles Krulak's 1998 White Letter "Sustaining the Transformation".)

Q19. How well does this unit sustain the *Transformation*?

- ☐ Extremely well
- ☐ Very well
- ☐ Moderately well
- ☐ Slightly well
- ☐ Not well at all

Photo by USMC/Marine Corps Lance Cpl. Tanner D. Lambert.

CHAPTER FIVE

SIGNIFICANCE, STUDY LIMITATIONS, FUTURE RESEARCH, AND FINAL THOUGHTS

This chapter explains the significance of the study, acknowledges a few limitations of the study, presents recommendations for future research, and provides some final thoughts. This research project has illustrated common categories and key themes through insights from eight CMC on command climate and its influence on the ethical behavior of the Marines in the organization.

Significance of the Study

Together with the literature review, the results of interviews with the eight CMC provide significant insights into the importance of command climate and the commander's ability to influence the ethical behavior of Marines in the organization in three ways. First, the insights from the CMC provide a unique perspective from an elite, senior group of leaders that is not often discussed in the literature focused on the area of leadership climate.

A second significant aspect of this study is that the Marine Corps prides itself on producing superior leaders at all levels and constantly strives for improvement. This study adds to the existing body of research and knowledge that the Marine Corps has accumulated to educate and train its leaders to be the most competent, committed, and fair leaders of virtue and character in both combat and in garrison.

The CMC suggested that the command climate established by the commanding officer does influence the ethical behavior of Marines. This impacts the Marine Corps' warfighting effectiveness, as well as its reputation as seen from the eyes of the American public. The study offers Marine leaders the opportunity to explore the detailed perspectives from the former CMC and the current Commandant of the Marine Corps, so that they might use this study as a heuristic to better examine their own command climates and make the necessary adjustments to promote the success of the individual Marine, the unit, and the Marine Corps as an institution.

Finally, the third significant aspect of this study is that it can benefit other leaders of organizations outside of the Marine Corps. Many of the CMC went on to other professions upon retirement, and applied the same leadership techniques they used when they established their command climates as leaders in the Marine Corps. For example, General Krulak became the President of Birmingham Southern University. He utilized many of the same leadership principles he used to establish the right command climate for the Marine Corps. Another example is General Jones, who went into business with Jones Group International, a consulting firm with clients all over the world.

This book provides tips and suggestions on how to establish a command climate, what a good command climate looks like, and how to develop and maintain the right command climate that will influence the ethical behavior of the people throughout the organization. This study of ethical command climate has relevance for leaders within any large organization, not just the United States Marine Corps. The research provided a recommended definition of command climate that any leader can apply to their organization. Further, the command climate survey on page 151 of this volume can be used to provide leaders feedback on their performance in setting the right organizational climate to promote ethical behavior and success for their organization.

Study Limitations

Even though this study offers a unique insight into the subject of climate from the perspectives of the CMC, there were some limitations. The first limitation is the demographics of the CMC, themselves. All are white, male, and, with one exception, infantry officers by background (General Amos was a naval aviator). In addition to this, one could argue that a former senior leader of the United States Marine Corps, a four-star general with typically 40 years or more of service, could be out-of-touch with a twenty-five-year-old Marine (current average age of a Marine) who, for example, uses social media to communicate. Although with age comes wisdom, the disparity in rank, position, gender, and racial bias coupled with age could tend to limit the validity of the findings if the sole objective was to develop insights based on what types of climate issues are only relevant to younger people.

The second limitation in this study centers on my positionality. As mentioned previously in the section covering methodology, my personal experiences and subjectivity could have skewed or shaped the study. However, subjectivity can be an asset rather than a liability if the researcher recognizes their subjectivity throughout the course of the research.[341] My personal background and experience as a commissioned Marine Corps officer with 28 years of experience became more of an asset than a liability in knowing what questions to ask the generals, and in gaining access to conduct the interviews. Nevertheless, my biases and subjectivity had to be monitored throughout the research process to ensure these areas were not an impediment during the research gathering, analyzing, and writing stages of the study.

The final limitation to the study was the issue of generalizability. The study's findings are not generalizable in the traditional scientific sense, because the use of a purposeful sampling strategy produces an unsaid

341. Peshkin, A. "In Search of Subjectivity-One's Own." *Educational Researcher* (October 1988): 17-21.

set of constraints when attempting to apply the findings to other population groups. As Donmoyer[342] points out, it is difficult to generalize from studies whose questions are focused on meaning and perspective, which was the case in this study. Nonetheless, the study is still valuable.

Future Research

Future research should examine the same subject from the perspectives of more junior leaders, such as company commanders. This study did not explore command climate from the perspectives of female leaders at any level, which should be studied based on the recent changes to Department of Defense policy regarding women serving in combat arms positions. Future research should also explore command climate and ethical behavior from the perspectives of enlisted leaders to gather their valuable insight, so as to inform current and future commanding officers as they establish their command climates.

Final Thoughts

The phase one review of the academic curriculum taught to both enlisted Marines and officers within the Marine Corps' formal schools, coupled with the review of the documents within the Marine Corps' doctrinal publications on leadership, demonstrated that the Marine Corps puts a heavy emphasis on character development and reinforces the Marine Corps' core values of honor, courage, and commitment to its Marines. Phase two interviews with the CMC revealed many initiatives from these senior leaders—such as the Crucible, The Reawakening Campaign, and the MCMAP—designed to assist commanders with their command climates and to sustain the Transformation. The interviews also revealed personal stories of how each Commandant's background and upbringing, as well as their observations of both good and bad leaders while serving in the Marine Corps, influenced their approach to command climate.

342. Donmoyer, R. "Generalizability and the Single-Case Study." In *Qualitative Research in Education: The Debate Continues*, Ed. by E. Eisner and A. Peshkin, 25. New York, NY: Teachers College Press, 1990.

And yet, despite the attention to character development and values, the Corps still suffers from breakdowns in ethical behavior by a small percentage of Marines who choose to take the wrong path. Some of the CMC discussed the problem of egos getting in the way of the command's focus. Others mentioned breakdowns due to human frailty, temptation, and exercising small breaches of integrity that developed into bigger problems over time. Finally, a few of the senior leaders discussed the issue of substance abuse, especially the abuse of alcohol, as contributing factors to ethical problems. These areas contribute to breakdowns in discipline and poor ethical decision making. As the results from this study suggest, the commanding officer sets the tone for the entire organization. The commander must live the core values and be seen by the followers as setting the right example.

The command climate must set the conditions for success of the individual Marine, the unit, and the Institution. The commander's climate must reinforce the core values and high standards on a routine basis, using ethical discussions led by junior leaders such as NCOs and junior officers. Finally, the command climate must promote the fact that America wants a Marine Corps and that it is the responsibility of the individual Marine to uphold the legacy of all the Marines who have gone before them. When a Marine violates the Marine Corps' standards and values, they let down not only their fellow Marines, but the entire Marine Corps as an institution. It is this area of "intangibles" where the commanding officer must focus.

It was truly an honor and a privilege to be able to interview such superb and gifted senior leaders of the Marine Corps that I love. The Marine Corps prides itself on leadership. This rang true during every interview with each Commandant. Upon completion of the interviews, I realized how fortunate I was to have served on active duty during the period when these leaders were serving in the Marine Corps, as either the Commandant of the Marine Corps or as general officers. The Marine Corps continues to be the most lethal and capable fighting force on the planet, but, equally as important, the Marine Corps continues to be an organization that has never let the American people down. It is truly an institution that is good for the country.

APPENDIX A

Research Methods

Methodology

This research project examined the role command climate plays in influencing the ethical behavior within a Marine Corps command. Two research questions guided the study:

1. How do the CMC describe a command climate that encourages ethical behavior among Marines?

2. How does a commanding officer develop a command climate that promotes ethical behavior?

This Appendix begins with a discussion of the research design and the two-phased operational approach that was employed in the study. The next section covers the research sites and participants followed by a discussion of the data collection methods. The discussion then turns to the analysis and the coding strategy used for the study.

A Case Study/Cross Case Analysis Design

To explore the research questions this study employed a case study/cross-case analysis design. A case study can stand on its own and provide the researcher with a detailed story from a participant, organization, or event.[343] In this case, the focus of the study and unit of analysis was the CMC and their perceptions on command climate, and how command climate influences ethical behavior within a Marine Corps command. The results of the analysis (patterns) across interview questions were

343. Patton, M. *Qualitative Research, and Evaluation Methods*. Thousand Oaks, CA: SAGE Publications Ltd., 2015; Yin, R. *Case Study Research: Design and Methods*. Thousand Oaks, CA: SAGE, 2003.

analyzed using the data collected from the open-ended interviews. The cross-case analysis was supplemented by two other data sources consisting of documents and observations to best "illuminate the inquiry."[344]

Merriam and Tisdell[345] discuss the value of triangulating emerging findings using interviews, document analysis, and observations to form solid conclusions. Patton contends that "triangulation, in whatever form, increases credibility and quality by countering the concern (or accusation) that a study's findings are simply an artifact of a single method, a single source, or a single investigator's blinders."[346] The research design of this study achieved such analytical rigor.

Research Overview—A Two-Phased Approach

The research was conducted in two phases. The first consisted of a visit to the MCWAR in Quantico, Virginia, to conduct a document review of Marine Corps doctrine and training manuals relative to command climate, interviews with key ethics training and education personnel, and observations of ethics discussions. The second phase consisted of personal interviews with seven former CMC and the current Commandant of the Marine Corps. I am grateful that each of these distinguished gentlemen allowed me to interview them.

The first phase was designed to accomplish three goals: 1) To enhance my understanding of command climate within the Marine Corps with an emphasis on current ethics and command climate pedagogy taught within its formal schools; 2) To interview two former senior commanders (colonels) who could help to provide context on command climate from the practitioner's perspective; and 3) To inform the development of the final interview guide that was to be used when interviewing the CMC.

344. Patton, M. *Qualitative Research, and Evaluation Methods*, p. 263.
345. Merriam, S. and Tisdell, E. *Qualitative Research, A Guide to Design and Implementation*. San Francisco, CA: Jossey-Bass, 2016.
346. Patton, M. *Qualitative Research, and Evaluation Methods*, p. 674.

Phase one began by conducting a review of current and past ethics and command climate curriculum as taught within the Marine Corps' officer development programs. To accomplish this task, documents were reviewed and discussions were conducted with key personnel at the MCWAR, the Expeditionary Warfare School, and the instructors at the LLI's Cornerstone Course for commanding officers. A review of these documents laid the foundation for the development of the interview guide and the preparatory work required prior to the interviews at Quantico.

The phase one work included personal interviews with two retired Marine Corps colonels. Each of these distinguished officers has commanded Marines and currently holds influential positions within the Marine Corps Association. Their perspectives on the command climate and its influence on ethical behavior were invaluable. Taken together, the review of the curriculum coupled with the interviews with former commanders helped to establish the necessary context to use in phase two. The phase one work assisted with the development of an effective interview guide that was used when discussing command climate and ethical behavior with the CMC.

The Director of Operations of the Marine Corps Association and Foundation, two ethics instructors from the LLI, and the publisher and editor of the Marine Corps' professional journal, *The Marine Corps Gazette*, were all interviewed during a round table discussion conducted at the LLI. These preliminary interviews were conducted to establish a context that was used to describe the historical disposition of the Marine Corps regarding ethical issues and ethics education during the period in which each Commandant served. This work also provided details on the current state of the Marine Corps' curriculum regarding command climate and shed light on what Marines are currently writing about regarding ethics and command climate. Overall, these interviews provided valuable insight into how the MCU is preparing Marine Corps commanding officers to create command climates that effectively deal with ethical issues.

Dr. Paolo Tripodi, a professor at the LLI, contributed to the data collection effort. Dr. Tripodi teaches ethics and leadership to numerous Marines throughout the Marine Corps. He provided insight on what is currently being taught to the students and future commanders assigned as students to the MCWAR. He also provided historical context on the command climate education that has been covered in the past, which supported the interviews and data collection efforts with the CMC. Lieutenant Colonel Derek Snell offered insight into the Marine Corps' commanders course called Cornerstone, designed to prepare board-selected officers to take command.

Colonel Tim Mundy, United States Marine Corps (Retired), is the Director of Foundation Operations at the Marine Corps Association and Foundation and a member of the MCU's recently-formed ethics and leadership working group. As a former commanding officer and a recently retired colonel with 30 years of experience, he was able to provide perspective on the subject matter and insight on the interview guide questions that were to be used to interview the CMC.

Colonel Chris Woodbridge, United States Marine Corps (Retired), the editor and publisher of the Marines' professional journal, the *Marine Corps Gazette*, was interviewed in Quantico. He provided unique insight into the ethical issues that Marines are currently writing about in their professional journal regarding command climate and ethical behavior.

Another set of preliminary interviews was conducted as part of a pilot study during 2016-17 and similarly informed the interview protocol for the larger study with the CMC. The interviews were conducted with seven commanding officers at the Marine Corps Recruit Depot, San Diego (two colonels, one lieutenant colonel, two majors, and two captains). The pilot study was designed to confirm the potential significance of the proposed study, to inform the study's draft interview protocol, to gather insights and perspectives from active-duty commanding officers responding to the interview guide questions, and to conduct observations that could contribute to the future study.

The pilot study found that commanders rely heavily on their background and experiences more than their formal military ethics education to make ethical decisions, that ethical programs must be regularly reinforced to establish and maintain an ethical climate, that training must be improved for future commanding officers on how to establish a command climate, and that ethical development and trust at the lowest level is required for a command climate to work.

The second phase consisted of one-on-one interviews with seven former CMC as well as the current Commandant of the Marine Corps. These were: General Alfred Gray (29th Commandant, 1987-1991), General Charles Krulak (31st Commandant, 1995-1999); General James Jones (32nd Commandant, 1999-2003), General Michael Hagee (33rd Commandant, 2003-2006), General James Conway (34th Commandant, 2006-2010), General James Amos (35th Commandant, 2010-2014), General Joseph Dunford (36th Commandant, 2014-2015), and General Robert Neller (37th Commandant, 2015-2019).

Patton advises that "the logic and power of purposeful qualitative sampling derives from the emphasis on the in-depth understanding of specific cases: *information-rich cases.* Information-rich cases are those from which one can learn a great deal about issues of central importance to the purpose of the research; thus, *purposeful sampling.*"[347]

Each interview was treated as an individual case. After the case descriptions were assembled and analyzed, the cases were compared with each other. The data collected informed the approach to the overarching concern of this study: What role does command climate play in influencing the ethical behavior within a Marine Corps command? As well, it served to answer the two research questions: 1) how do the CMC describe a command climate that encourages ethical behavior among Marines? and 2) how does a commanding officer develop a command climate that promotes ethical behavior?

347. *Ibid*, p. 53.

A case study/cross-case comparison design was chosen for two reasons. First, case studies were used because they provide an in-depth description and analysis of a "contemporary phenomenon (the "case") within its real-life context."[348] The case studies offered a unit of study to gain a deeper insight into a specific issue (command climate) and to better understand the context and other factors that influence the issue.

The case study/cross-case comparison design revealed individual respondent priorities, experiences, and approaches while also providing a means for comparison between respondents. The design also generated insight from senior leaders who have served as Commandant during different presidential administrations, and during very different social/political time periods; for example, the era before 9/11, or prior to the advent of social media. This study included interviews with eight senior leaders that facilitated thick descriptions of command climate and how it influences ethical behavior. The interviewees provided a unique perspective on the Marine Corps' approach to command climate and ethics over the past 30 years. This is of paramount importance since ethical concerns involving Marines have risen to the attention of the Commandant of the Marine Corps, the Chairman of the Joint Chiefs of Staff (currently General Joseph Dunford, a former Commandant who was interviewed for this study), the other Service Chiefs (senior leaders of the other Services of the Armed Forces), the Secretary of Defense, the Congress, the President of the United States, and the American people.

These leaders also discussed how to create command climates that influence ethical behavior based on their experiences, education, and training. A case study/cross-case comparison design accommodated the diversity

348. Merriam, S. and Tisdell, E. *Qualitative Research, A Guide to Design and Implementation*, p. 37.

within the commanding general population and the different political and social eras of their time better than most other research designs.[349]

Participants and Research Sites

The participants were recruited through an email invitation sent to the senior leaders help desk at Headquarters Marine Corps. It included a note summarizing the research project and why I was asking for their participation.[350] A similar note was sent to the participants at Quantico along with their respective chains of command as a courtesy, and to ensure senior officers were aware of the request for participation.[351] The CMC were interviewed at their place of business, residence, or—as in one case—by phone.

Data Collection Procedures

Data was collected from three sources. First, one-on-one personal interviews with the seven former CMC and the current Commandant. Second, interviews and group discussions with key personnel at Quantico, with participants responding to questions from an open-ended, semi-structured interview guide.[352] Third, document reviews of various Marine Corps orders, letters, instructions, doctrinal publications, handouts, and curriculum taught at formal schools for both officer and enlisted personnel. In addition, observations from the pilot study were incorporated to inform the analysis.

349. *Ibid.*
350. See Appendix B.
351. Glesne, C. *Becoming Qualitative Researchers, An Introduction*, New York, NY: Pearson, 2016.
352. See Appendix C.

Interviews

Open-ended, semi-structured interviews were the primary data collection mechanism used in this research project. This approach was selected because, as Merriam and Tisdell[353] argue, semi-structured, open-ended interviews are well suited for addressing key topics while allowing for unplanned questions and topics to emerge during fieldwork.

The interview guide approach was used in this study to provide structure given the time limitations normally associated with officers in such high positions with a multitude of other commitments. Patton[354] suggests that the interview guide provides a comprehensive methodology that works well when there are time boundaries or other constraints associated with the interview. Interviews lasted between 60-90 minutes on average, with the longest lasting three hours. All interviews were audio-recorded and then transcribed verbatim for analysis.

During the interviews, I was able to take notes, and afterwards prepared an analytical memo for each.[355] The same process was used for the discussions with the key personnel at Quantico prior to the interviews with the CMC. Follow-up interviews were not necessary after the initial interviews were completed. However, I mailed each Commandant his respective section to validate the information therein. I also sent my proposed definition of command climate to one Commandant seeking his thoughts and guidance.

After each Commandant reviewed his section, he emailed me minor corrections, or merely responded stating the section accurately reflected the thoughts he provided during the interview. All corrections and suggestions from the CMC were included in the final version of the study. There were no significant alterations required during the member check phase of the writing process.

353. Merriam, S. and Tisdell, E. *Qualitative Research, A Guide to Design and Implementation.*
354. Patton, M. *Qualitative Research, and Evaluation Methods.*
355. Merriam, S. and Tisdell, E. *Qualitative Research, A Guide to Design and Implementation.*

Document Analysis

A document analysis was conducted not only to understand what the Marine Corps has been teaching regarding ethics instruction but also to analyze what is being taught on command climate with an emphasis on content analysis regarding ethical climate.[356] The document review provided some historical context that informed the interviews with the CMC.

A review of Marine Corps doctrine and training publications was conducted prior to this study as part of the literature review. The literature review provides an overview of ethics education at the Marine Corps Senior Service College. Also, various doctrinal manuals were reviewed, including MCWP 6-11, *Leading Marines*, 2014; MCRP 6-11, *Marine Corps Values: A User's Guide for Discussion Leaders*, 2014; and MCWP 5-1, *Marine Corps Planning Process*, 2010. These manuals helped to provide a greater understanding of the Marine Corp's position on ethics and the emphasis on its core values.

During the document review portion of the study, the command climate instruction offered at the Cornerstone Course for commanders was a primary focus. Although not the overall focus of this study, other ethics instruction provided to the junior officers at the Expeditionary Warfare School and the senior officers who attend the war college were reviewed for context. These additional areas of ethics curriculum were examined in search of information that addresses command climate and how it influences the ethical behavior within a Marine Corps command.

Observations

One observation session was conducted during the pilot study prior to the research project. A group of approximately 30 Marines who

356. Merriam, S. and Tisdell, E. *Qualitative Research, A Guide to Design and Implementation.*

participated in an ethics discussion that was directed by their commanding officer were observed for approximately one hour. This discussion was a routine part of the commanding officer's command climate. The observation occurred at the Marine Corps Recruit Depot in San Diego, California and was led by junior non-commissioned officers (NCOs, corporals and sergeants). The observation informed the study by underscoring the importance of reinforcing ethical training within a Marine Corps command and having the instruction led by junior leadership. The observation helped triangulate some of the findings that emerged from the interviews with the CMC.[357] The observed discussions focused on ethics and the chain of command (whom Marines should go to when they have questions or issues). The unit's commanding officer believed these recurring discussions have contributed to a reduction in ethics-related problems within his command.

Data Analysis and Coding Strategy

The theoretical framework underpinning the data analysis and coding strategy for this study was grounded theory.[358] The study utilized codes that were applied to data through a series of cumulative coding cycles, which produced categories and themes: "A theme can be an *outcome* of coding, categorization, or analytic reflection."[359] These codes are discussed in greater detail below.

The research employed both narrative and document analyses with the goal of creating "translucent windows into cultural and social meanings when understood and analyzed as narratives."[360] The narrative aspect of the qualitative inquiry is the story. The analysis occurs when the story is recorded through interviews, transcribed, and scanned for patterns that assist the researcher in learning more about the individual and the society and culture in a general sense.[361] This research project treated

357. Merriam, S. and Tisdell, E. *Qualitative Research, A Guide to Design and Implementation.*
358. Glesne, C. *Becoming Qualitative Researchers, An Introduction*; Merriam, S. and Tisdell, E. *Qualitative Research, A Guide to Design and Implementation.*
359. Saldana, J. *The Coding Manual for Qualitative Researchers.* Thousand Oaks, CA: Sage Publications, Inc., 2016, p. 15.
360. Patton, M. *Qualitative Research, and Evaluation Methods*, p. 128.
361. *Ibid.*

each Commandant's story as data and then analyzed the narratives. This required interpreting the stories from the interviews, putting the stories into context, and comparing them with one another.

Glesne[362] argues that narrative analysis can foster understanding of how interviewees construct meaning based on their experiences. Stories told by participants may include narratives linked to their cultural and political context. The interviewer can then examine how the participant correlates his or her experiences and circumstances to make meaning. This study gathered narratives from participant responses to questions from the interview guide to develop codes, categories, and themes.

The narratives from the interviewees coupled with their responses to interview guide questions were used to examine and deepen the collective understanding of a commanding officer's approach to establishing a climate that influences the ethical behavior of the Marines in the command. This process combined two qualitative methodologies—narratives and grounded theory.[363] As Merriam and Tisdell point out, "one could build grounded theory within a case study, or present a person's 'story,' hence combining narrative with a case study."[364]

Polkinghorne uses the term "narrative configuration"[365] to describe the process whereby events are drawn together and integrated into a temporary whole. The data collected from the individual stories and the responses to the interview guide questions were used in the study and coded to facilitate the analysis. The individual case studies were constructed using narrative analysis to facilitate a greater understanding of events from the participant's past and the factors or elements of an organizational climate that influenced ethical behavior.

The first interviews with key personnel at Quantico were used to help inform the analysis and to establish context on command climate from

362. Glesne, C. *Becoming Qualitative Researchers, An Introduction.*
363. *Ibid.*
364. Merriam, S. and Tisdell, E. *Qualitative Research, A Guide to Design and Implementation*, p. 39
365. Polkinghorne, D. "Narrative Configuration in Qualitative Analysis." Qualitative Studies in Education 8 no. 1 (1995): 5-23.

a recent practitioner's point of view. These officers also assisted in the development of the interview guide prior to the Commandant interviews. Once all eight cases were developed (seven former CMC, and the current Commandant), the participants' responses to the interview questions were compared to look for differences and similarities.[366] Coding was required to organize the data into four general categories: setting the example, open communication, core values, and accountability and responsibility.[367] The coding method was divided into two main sections: the first cycle and second cycle coding methods.[368]

The first cycle of coding employed in vivo coding (based on the direct language used by the participants).[369] Axial coding was used as the second cycle coding method to develop subcategories through the disaggregation of core themes (categories and concepts) via a combination of inductive and deductive reasoning.[370] The manual method of coding was used during this process to organize and manage the data. In addition to the coding process, analytical memo writing was used during the process to capture reflections from the interviews, document reviews, and the observation to make connections between categories, themes, and recommendations.[371]

366. Glesne, C. *Becoming Qualitative Researchers, An Introduction.*
367. See Appendix D.
368. Saldana, J. *The Coding Manual for Qualitative Researchers.*
369. *Ibid.*
370. *Ibid.*
371. *Ibid.*

APPENDIX B

DEAR COMMANDANTS LETTER

Dear Commandants,

I am honored and humbled to send you this note requesting to interview you about command climate. My name is Brian Kerl, and I retired in 2013 as a colonel of Marines. Since my retirement, I have attended night school as a student at the University of San Diego's School of Leadership and Education Sciences working on my Ph.D. in leadership while running the MCJROTC program at Oceanside High School. I am assisting the Marine Corps University with research on command climate. I believe your contribution to my study would be invaluable. I wish to interview you before the end of the year for about an hour or so.

My study will examine a Marine Corps commanding officer's approach to his or her command climate and how it influences the ethical behavior of the Marines. What training do we need to provide our commanders to establish an ethical climate? What does an ethical climate look like? How does a commander establish an ethical climate?

Dr. Paolo Tripodi (LLI) will assist me throughout the research phase of the process and has agreed to be on my dissertation committee. I believe that my research will be able to contribute to the work already initiated by the Lejeune Leadership Institute. I would appreciate your willingness to be interviewed, and I can be reached at 513-335-2303, or email: bdkerl3@gmail.com. Thank you for your consideration.

Very Respectfully,

Brian Kerl

Photo by dovapi.

APPENDIX C

INTERVIEW GUIDE

INTRODUCTION

I have given you a consent form that asks for your permission to participate in this study. I will give you a few minutes to review the information on the consent form and confirm that you still are interested in participating.

Thank you for coming in today. Before we begin, I want to introduce myself and explain why I am conducting this study. My name is Brian Kerl, and I am currently a doctoral student in the School of Leadership and Education Sciences at the University of San Diego. I am conducting this study as part of my PhD program in leadership to fill a gap in leadership knowledge at the Marine Corps University in Quantico, and to offer a study to the university that will add to its existing literature to assist Marine Corps commanding officers in their approach to establishing an organizational climate that positively influences ethical behavior.

The purpose of this study is to better understand how command climate influences ethical behavior in a Marine Corps unit. My goal is to begin to explore the connection between these two phenomena using the theoretical framework for describing ethical leadership, ethical climate, and the ethical underpinnings associated with ECT. I've developed a set of questions to guide our conversation. Please feel free to share as much as you are comfortable sharing and understand that you are free to choose not to participate in all or any part of the study.

I would like your permission to audio-record the interview and to take notes during our conversation. I will be the only person who will use the notes and the audio recording to ensure I have accurately recorded your responses. Is this all right? Do you have any questions?

RESEARCH QUESTION

What role does command climate play in influencing the ethical behavior within a Marine Corps command?

1. First, please introduce yourself and briefly describe your military background.

2. Comment on the key events throughout your life and career that have had the greatest influence on your approach to your command climate?

3. While you were Commandant, how did the Marine Corps approach command climate? How does command climate influence ethical behavior and why is it important?

SUPPORTING QUESTION #1

How do the Commandants describe a command climate that encourages ethical behavior among Marines?

4. Describe what a command climate looks like that encourages ethical behavior? What are the key elements or components of a command climate that encourages ethical behavior?

5. Based on today's society and the challenges facing Marine commanding officers; what should the Marine Corps teach its commanders relative to climate and its ability to influence ethical behavior?

SUPPORTING QUESTION #2

How does a commanding officer develop a command climate that promotes ethical behavior?

6. What steps should a commander take to establish a command climate that promotes ethical behavior?

7. How does the commander know the climate is working? Can you offer some examples?

8. What direction if any, did you provide training and education command regarding the preparation required for Marine commanders and the climates they are expected to set?

Photo by kvkirillov.

APPENDIX D

Coding Cycles and Codes

1st Cycle Coding Categories	Codes (32)	2nd Cycle Coding Categories	Codes (24)
Leading Others	Role Model, Humility, Commander's Philosophy, Guidance, Ethical Lapse, Transparency, Families, Mission, Shaping, Cohesion, Influence, Environment, Maneuver Warfare Concept, Little Things	Setting the Example "Living It"	Selflessness, Trust, Servant Leadership, the Commander, Command Climate, The Transformation
Talking to People	Listening, Observations, Relationships, Feedback, Visibility, Distractions, Trust	Open Communication	Open Door Policy, Critiques, Face-to-Face, Expectations, Communicator-in-Chief, Being Human
Beliefs/Values	Ethical Decision Making, Teacher-Scholar, Ethical Dilemmas, Character, Standards, Values, Institution	Core Values	Dignity, Intangibles, Balanced Excellence, High Standards, Socratic method, lowest level, Reinforcement, Hearts and Souls,
Discipline	Self and others, Rules and Regulations, Expectations, Accountability	Accountability and Responsibility	Discipline, Protecting the Institution, CMC, Commanders,

Photo by: ruskpp.

APPENDIX E

Official Biographies of the Commandants

GENERAL ALFRED M. GRAY Jr.

General Al Gray, USMC (Ret) serves as Chairman of the Board of Regents; Member of the Board of Directors; and Senior Fellow of the Potomac Institute of Policy Studies (PIPS). The PIPS serves as a non-partisan, not-for-profit policy research institute that provides an academic forum for the study of key national security, science and technology, and related policy issues.

General Gray has served as Board Chairman and CEO for several public and private companies and has consulted to United States and international industry and government. General Gray's other duties have included service on the Defense Science Board; the Defense Special Operations Advisory Group; the National Security Agency Science Advisory Board; the National Reconnaissance Office Gold Team; the Defense Operations Support Office Advisory Team; and as Director of the Advanced Concept Demonstration Technology for Combat in the Littorals. Consistent with his interest in education and helping servicemen and women, he is trustee Emeritus of Norwich University, past trustee on Monmouth University, past member of the National Defense University Board of Visitors and is Chairman Emeritus of American Military University. He is Chairman Emeritus of The Injured Marine Semper Fi Fund and the America Fund, having served over ten years, to help take care of wounded veterans and their families.

General Gray currently serves as Chancellor of The Marine Military Academy, Chairman of the US Marine Youth Foundation, and as the Board of Trustee Chairman at the American Public University System.

In 1991, Al Gray retired after 41 years of service to the United States Marine Corps. From 1987-1991, General Gray served as a member of the Joint Chiefs of Staff, was the 29th Commandant of the Marine Corps, and was advisor to both Presidents Reagan and George H. W. Bush. As Commandant, he instituted and published a Warfighting Philosophy for Marines based on the Maneuver Warfare Thought Process. General Gray developed and implemented a new long-range strategic planning process for the Marine Corps, established the Marine Corps University, and implemented other longstanding changes, such as ensuring that every Marine is a rifleman first and that the Marine Corps was Special Operations Capable.

General Gray enlisted in the Marine Corps in 1950 and achieved the rank of Sergeant while serving in amphibious reconnaissance with Fleet Marine Force, Pacific aboard the submarine USS *Perch* (ASSP-313). He was commissioned a Second Lieutenant in 1952. In the early years, he held extensive assignments overseas in the Far East and Southeast Asia in infantry, artillery, intelligence, communications and special operations. He has held every infantry command assignment from Platoon Commander through Division Commander and has commanded every Marine Air Ground Task Force from Marine Corps Expeditionary Unit to Marine Expedition Force.

GENERAL CHARLES C. KRULAK

General Charles C. Krulak, 31st Commandant of the Marine Corps, was born on 4 March 1942 in Quantico, Virginia. After graduating from The Phillips Exeter Academy in Exeter, New Hampshire, General Krulak attended the U.S. Naval Academy. He graduated from the academy in 1964 with a Bachelor of Science degree in Engineering. General Krulak also holds a Master of Science degree in Labor Relations from The George Washington University (1973). He is a graduate of the Amphibious Warfare School (1968); the Army Command and General Staff College (1976); and the National War College (1982).

Since his commissioning and graduation from The Basic School at Quantico, General Krulak has held a variety of command and staff positions. He commanded a platoon and two rifle companies during his two tours of duty in Vietnam; Commanding Officer, Special Training Branch, and Recruit Series, MCRD, San Diego, California (1966-1968); Commanding Officer Counter-Guerilla Warfare School, NTA, Okinawa (1970); Company Officer, U.S. Naval Academy (1970-1973); Commanding Officer, Marine Barracks, Naval Air Station, North Island, California (1973-1976); and Commanding Officer 3d Battalion, 3d Marines (1983-1985).

General Krulak's staff assignments include: S-3 of the 2d Battalion, 9th Marines (1977-1978); Chief of the Combat Arms Monitor Section, Headquarters Marine Corps, Washington, D.C. (1978-1979); Executive Assistant to the Director of Personnel Management, Headquarters Marine Corps (1979-1981); Plans Officer at FMF Pacific (1982-1983); Executive Officer, 3d Marines, 1st MEB; Assistant Chief of Staff for Maritime Prepositioning Ships, 1st MEB; Assistant Chief of Staff, G-3, 1st MEB; and the Military Assistant to the Assistant Secretary of Defense for Command, Control, Communications and Intelligence, Office of the Secretary of Defense.

He was assigned duty as the Deputy Director of the White House Military Office in September 1987. While serving in this capacity, he was selected for promotion to brigadier general in November 1988. He was advanced to that grade on 5 June 1989, and assigned duties as the Commanding General, 10th Marine Expeditionary Brigade/Assistant Division Commander, 2d Marine Division, FMF, Atlantic, Camp Lejeune, North Carolina, on 10 July 1989. On 1 June 1990 he assumed duties as the Commanding General, 2d Force Service Support Group/Commanding General 6th Marine Expeditionary Brigade, FMF, Atlantic, Camp Lejeune. He served in this capacity until 12 July 1991, and was assigned duty as Assistant Deputy Chief of Staff for Manpower and Reserve Affairs (Personnel Management/Personnel Procurement), Headquarters Marine Corps on 5 August 1991. He was advanced to major general on 20 March 1992. General Krulak was assigned as Commanding General, Marine Corps Combat Development Command, Quantico, Virginia on 24 August 1992, and was promoted to lieutenant general on 1 September 1992. On 22 July 1994, he was assigned as Commander of Marine Forces Pacific/Commanding General, Fleet Marine Forces Pacific, headquartered at Camp H.M. Smith, Hawaii.

General Krulak was promoted to general 29 June 1995, and assumed duties as the 31st Commandant of the Marine Corps on 1 July 1995. He retired from the Marine Corps on 30 June 1999.

During his military career, General Krulak was awarded the Defense Distinguished Service Medal; Distinguished Service Medal; Silver Star Medal; Bronze Star Medal with Combat "V" and two gold stars; Purple Heart with gold star; Meritorious Service Medal; Navy and Marine Corps Commendation Medal; Combat Action Ribbon; Presidential Unit Citation with bronze star; National Defense Service Medal with one bronze star; Vietnam Service Medal with silver star and two bronze stars; Southwest Asia Service Medal with two bronze stars, Sea Service Deployment Ribbon; Republic of Vietnam Cross of Gallantry; Republic of Vietnam Meritorious Unit Citation (Gallantry Cross Color); Republic of Vietnam Meritorious

Unit Citation (Civil Actions Color, 1st Class); the Republic of Vietnam Campaign Medal; and the Kuwait Liberation Medal.

Upon his retirement from the Marine Corps, General Krulak served as a member of the Board of Directors of Conoco and then Conoco Phillips, Union Pacific Railroad, Phelps Dodge, Freeport McMoran Copper and Gold, the Cleveland Browns, Aston Villa FC, a premier soccer club in the United Kingdom and was the Chairman of the Naval Academy Board of Visitors.

General Krulak was introduced as the 13th President of Birmingham-Southern College on March 21, 2011 and retired as of June 30, 2015.

In addition, General Krulak is an advisor to the Center for Naval Analysis, worked with Human Rights First to disrupt the business of modern day slavery and human trafficking, is a member of the Council on Foreign Relations and is on the Board of Directors for the U.S.-Israel Education Association and the Community Foundation of Greater Birmingham.

GENERAL JAMES L. JONES

As President of Jones Group International (JGI) and one of America's leading authorities on foreign policy, energy security, and national security, General Jones provides JGI clients with strategies for navigating the complex nexus of business, international affairs, and public policy.

General Jones was appointed by President Barack Obama as National Security Advisor to the President on January 20, 2009. During his tenure in the White House, General Jones served as a trusted Presidential advisor, represented the President as an envoy to American allies and partners, provided steady leadership during times of conflict, and oversaw an expansion of responsibilities of the National Security Council to include cyber security, homeland security, and strategic foresight.

Upon General Jones' retirement, President Obama said, "Jim has always been a steady voice in Situation Room sessions, daily briefings, and with meetings with foreign leaders...the American people owe the General a debt for making the nation safer."

General Jones came to the White House from the private sector, where he served as the President and CEO of the U.S. Chamber of Commerce's Institute for 21st Century Energy. At the Chamber, General Jones worked to unite energy consumers and producers in pursuit of common goals—to increase U.S. energy supply and improve infrastructure, to advance international cooperation on energy issues, to protect national energy security, and to promote a better understanding of changes to the global climate and its effects on the environment.

While leading the Chamber's energy work, General Jones also served in the George W. Bush administration as the State Department's Special Envoy for Middle East Regional Security. In this capacity, he worked with Israeli and Palestinian officials in furthering the Peace Process, focused on strengthening security for both parties to the conflict.

General Jones retired from the U.S. Marine Corps in February 2007 after a distinguished forty-year career. From July 1999 to January 2003, General Jones served as the 32nd Commandant of the United States Marine Corps, the most senior position in the Corps.

In 2003, General Jones was nominated to serve as Commander, United States European Command and Supreme Allied Commander Europe. As Commander of U.S. European Command, General Jones' area of responsibility included 92 countries from Europe, Eurasia, and Africa. In his capacity as Commander of all NATO forces, General Jones led the Alliance into overall command of the International Security Assistance Force in Afghanistan, the largest combat mission in NATO's history. General Jones also oversaw the military integration of the Alliance as it expanded from 19 to 26 members, and he advocated that NATO take on energy security and defense of critical infrastructure as a core mission.

General Jones spent the formative years of his youth living in France, where he became fluent in French and developed a global perspective. General Jones graduated from the Georgetown University School of Foreign Service and was commissioned as a Second Lieutenant in the Marine Corps in January 1967. He served as a rifle platoon and Company Commander in Vietnam, where he earned a Silver Star. On returning to the U.S., he attended the Amphibious Warfare School in 1973 and the National War College in 1985 and served as Marine Corps Liaison Officer to the U.S. Senate. In addition, he served as Commanding Officer of the 24th Marine Expeditionary Unit in Northern Iraq and Turkey on Operation Provide Comfort; Chief of Staff, Joint Task Force Provide Promise, for operations in Bosnia-Herzegovina and Macedonia; and Commanding General, 2nd Marine Division, Marine Forces Atlantic. He also served as Military Assistant to the Secretary of Defense from 1997-99.

Upon leaving the White House in 2010, General Jones founded Jones Group International. General Jones and the JGI team leverage their unmatched expertise, access, and credibility to assist clients in matters of energy security, national and international security, market access and trade promotion, and strategic leadership.

GENERAL MICHAEL W. HAGEE

General Michael W. Hagee retired on 1 January 2007 after serving as the 33rd Commandant of the Marine Corps from January 2003 through November 2006.

General Hagee graduated with distinction from the U.S. Naval Academy in 1968 with a Bachelor of Science in Engineering. He also holds a Master of Science in Electrical Engineering from the U.S. Naval Postgraduate School and a Master of Arts in National Security and Strategic Studies from the Naval War College. He is a graduate of the Command and Staff College and the U.S. Naval War College.

General Hagee's command assignments include: Commanding Officer Company A, 1st Battalion, 9th Marines 1970; Platoon Commander, Company A and Commanding Officer Headquarters and Service Company, First Battalion, First Marines (1970-1971); Commanding Officer, Waikele-West Loch Guard Company (1974-1976); Commanding Officer, Pearl Harbor Guard Company (1976-1977); Commanding Officer, 1st Battalion, 8th Marines (1988-1990); Commanding Officer, 11th Marine Expeditionary Unit (Special Operations Capable) (1992-1993); Commanding General, 1st Marine Division (1998-1999); and Commanding General, I Marine Expeditionary Force (2000-2002).

General Hagee's staff assignments include: Communications-Electronics Officer, 1st Marine Air Command and Control Squadron (1971); Assistant Director, Telecommunications School (1972-1974); Training Officer, 3d Marine Division (1977-1978); Electrical Engineering Instructor, U.S. Naval Academy (1978-1981); Head, Officer Plans Section, Headquarters Marine Corps (1982-1986); Assistant Chief of Staff, G-1, 2d Marine Division (1987-1988); Executive Officer, 8th Marines (1988); Director Humanities and Social Science Division/Marine Corps Representative, U.S. Naval Academy (1990-1992); Liaison Officer to the U.S. Special Envoy to Somalia (1992-1993); Executive Assistant to the Assistant Commandant of the Marine Corps (1993-1994); Director, Character Development Division, United States Naval Academy (1994-1995); Senior Military Assistant to the Deputy Secretary of Defense, Washington, D.C.; Executive Assistant to the Director of Central Intelligence (1995-1996); Deputy Director of Operations, Headquarters, U.S. European Command (1996-1998); and Director Strategic Plans and Policy, U.S. Pacific Command (1999-2000).

GENERAL JAMES T. CONWAY

General James T. Conway retired from active duty in 2010 after serving our nation for 40 years. Since then, he has joined and consulted for numerous corporate and non-profit boards, including Textron Inc., Colt Defense and General Dynamics. He also co-chairs the Energy Security Leadership Council (Securing America's Future Energy), a non-partisan energy policy think tank. He is also a member of the board for VISLINK.

Prior to his retirement, General Conway served as the 34th Commandant of the U.S. Marine Corps for four years. As Commandant, he served as the senior uniformed Marine responsible for the organization, training, and equipping of over 250,000 active duty, reserve, and civilian Marines throughout the United States and overseas, as well as the management of the $30-$40 billion annual Marine Corps budget.

Prior to becoming Commandant, General Conway served on the Joint Chiefs of Staff as Senior Operations Officer (J-3) in the U.S. military, where he oversaw the war efforts in Iraq and Afghanistan. As a member of the Joint Chiefs of Staff, General Conway functioned as a military advisor to the Secretary of Defense, the National Security Council, and the president.

During his tenure, the Marine Corps experienced growth of 27,000 additional personnel, undertook major military construction efforts at bases and stations throughout the U.S., and integrated multiple next-generation weapons systems into its inventory. General Conway initiated new programs for the care of wounded warriors and brought Marine Corps family programs onto a war-time footing.

Among his numerous leadership positions, General Conway served as Commanding General of the Second Marine Division, and he was the commander of the First Marine Expeditionary Force (I MEF) during two combat tours in Iraq. He initially led the invasion of Iraq in 2003 with 90,000 troops under his command, including Marines, soldiers, sailors, and British forces. During his second tour he oversaw the first Battle of Fallujah. General Conway graduated with honors from the Basic School, the U.S. Army Infantry Officers Advanced Course, the Marine Corps Command and Staff College and the Air War College. He was born in Walnut Ridge, Arkansas and graduated from Southeast Missouri State University in 1969. He was commissioned as an infantry officer in 1970.

General Conway is married to the former Annette Drury of Ste Genevieve, MO. They have three children, all currently serving or affiliated with the Marine Corps. Colonel J. Brandon Conway is currently commanding the 12th Marine Corps Recruiting District at MCRD, San Diego and Colonel select A. Scott Conway is the former Battalion Commander of 3d Raider Battalion. Their daughter, Samantha, is married to Colonel select David B. Moore, former Commanding Officer of a Marine Cobra Squadron. General Conway and his wife have seven grandchildren, Clara, Amelia, James, Leif, Will, Zach, and Beckett.

GENERAL JAMES F. AMOS

General Amos retired as the 35th Commandant of the U.S. Marine Corps. He currently serves as Chairman of the Board of Directors of Lord Corporation, Strategic Advisor to President—Vision Technology Systems Inc.—ST Engineering, member of Huntington Ingalls Shipbuilding Strategic Advisory Panel, Trustee NOVANT Health Corp, member of Charlotte's Veterans Bridge Home Advisory Board, member Board of Advisors, Jewish Institute for National Security in America (JINSA), member Advisory Board—With Honor Fund—a nonpartisan veteran focused political action initiative, Founder, Windsock LLC, former Chairman of the Board of Directors of the Semper Fi Fund.

A native of the Pacific Northwest, General Amos joined the Marine Corps after college, entering pilot training in 1970. He flew fighter aircraft in numerous Marine squadrons throughout the early years of his career, culminating in command of an F/A-18 Hornet squadron onboard the nuclear carrier, USS *Theodore Roosevelt*, and later command of Marine Aircraft Group 31.

Following his selection to General in 1998, Amos' varied assignments found him in the Balkans with NATO forces during the Serbian/Kosovo conflict, in Iraq twice in command of the 3rd Marine Aircraft Wing, at Camp Lejeune NC in command of all east Coast Marine combat forces, finally, culminating with his assignment in 2010 as the 35th Commandant of the Marine Corps.

Completing 42 years of service to our nation and its Marines, General Amos retired from active duty in December 2014. He and his bride of 48 years remain active in public and private business ventures, veteran philanthropy work nationally, and in their hometown of Charlotte, NC.

GENERAL JOSEPH F. DUNFORD, Jr.

Photo By: Department of Defense

General Joseph F. Dunford, Jr. is the 19th Chairman of the Joint Chiefs of Staff, the nation's highest-ranking military officer, and the principal military advisor to the President, Secretary of Defense, and National Security Council.

Prior to becoming Chairman on October 1, 2015, General Dunford served as the 36th Commandant of the Marine Corps. He previously served as the Assistant Commandant of the Marine Corps from 2010 to 2012 and was Commander, International Security Assistance Force and United States Forces-Afghanistan from February 2013 to August 2014.

A native of Boston, Massachusetts, General Dunford graduated from Saint Michael's College and was commissioned in 1977. He has served as an infantry officer at all levels, to include command of 2nd Battalion, 6th Marines, and command of the 5th Marine Regiment during Operation IRAQI FREEDOM.

General Dunford also served as the Assistant Division Commander of the 1st Marine Division, Marine Corps Director of Operations, and Marine Corps Deputy Commandant for Plans, Policies and Operations. He commanded I Marine Expeditionary Force and served as the Commander, Marine Forces U.S. Central Command.

His Joint assignments include duty as the Executive Assistant to the Vice Chairman of the Joint Chiefs of Staff, Chief of the Global and Multilateral Affairs Division (J-5), and Vice Director for Operations on the Joint Staff (J-3).

A graduate of the U.S. Army Ranger School, Marine Corps Amphibious Warfare School, and the U.S. Army War College, General Dunford also earned master's degrees in Government from Georgetown University and in International Relations from the Fletcher School of Law and Diplomacy.

GENERAL ROBERT B. NELLER

General Robert B. Neller is the 37th Commandant of the United States Marine Corps. Prior to his current assignment, he served as the Commander, Marine Forces Command from July 2014 to September 2015 and Commander, Marine Forces Central Command from September 2012 to June 2014.

A native of East Lansing, Michigan, General Neller graduated from the University of Virginia and was commissioned in 1975. He has served as an infantry officer at all levels, including command of Marine Security Force Company Panama during *Operations Just Cause* and *Promote Liberty*; 3d Light Armored Infantry Battalion during *Operation Restore Hope*; 6th Marine Regiment; and 3d Marine Division. General Neller also served as Deputy Commanding General, I Marine Expeditionary Force (Forward) during *Operation Iraqi Freedom* (05-07); Assistant Division Commander for 1st and 2d Marine Divisions; and President of Marine Corps University. His Joint assignments include service in the Policy Division of Supreme Headquarters Allied Powers Europe (SHAPE) in Casteau, Belgium, and as the Director of Operations (J-3) of the Joint Staff in Washington, D.C. General Neller is a graduate of the Armor Officer Advanced Course, Marine Corps Command and Staff College, NATO Defense College, and the Armed Forces Staff College. He holds a Master's Degree in Human Resource Management from Pepperdine University.

Bibliography

Allen, C. "Ethics and Army Leadership: Climate Matters." *Parameters* 45 no. 1 (2015): 69-83.

Allison, G., and Zelikow, P. *Essence of Decision, Explaining the Cuban Missile Crisis.* New York, NY: Addison-Wesley Publishers Inc., 1999.

American Heritage. *Dictionary of the English Language, 5th edition.* Boston, MA: Houghton Mifflin Harcourt Publishing co., 2013.

Amos, J. "Command Climate.". White Letter No. 3-13. Headquarters Marine Corps, Washington, D.C., 2013a.

Amos, J. "Leadership and Conduct." White Letter No. 1-12. Headquarters Marine Corps, Washington, D.C., 2012a.

Amos, J. "Leadership." White Letter No. 3-12. Headquarters Marine Corps, Washington, D.C., 2012b.

Amos, J. *The Reawakening.* Headquarters Marine Corps, Washington, D.C., 2013b.

Amos, J. "Heritage and Values Brief [Power Point Slides]." Retrieved from: https://www.dvidshub.net/video/157241/Marine-Corps-heritage-and-values.

Amos, J. "Commandant's Report to Congress on the Posture of the Marine Corps." Headquarters Marine Corps, Washington, D.C., 2014.

Bailey, J. "Interpersonal Skill Development: A Requirement for TBS." *Marine Corps Gazette* 100, no. 7 (2016): 10-13.

Barnes, C. and Doty, J. "What Does Contemporary Science Say About Ethical Leadership?" *Military Review* 90, no. 5 (2010): 90-93.

Bazerman, M., and Tenbrunsel, A. *Blind Spots: Why We Fail to Do what's Right and What to Do about it.* Princeton, NJ: Princeton University Press, 2011.

Behn, B. U.S. Air War College. "The Stakes are High: Ethics Education at US War Colleges." Graduation Thesis, Air University, 1-27. Maxwell AFB, AL: Air University Printing, 2016.

Bowen, D. and Ostroff, C. "Understanding HRM-firm performance linkages: The Role of "strength" of the HRM system." *Academy of Management Review* 29 (2004): 203-221.

Bryman, A., Collinson, D., Grint, K., Jackson, B., and Uhl-Bien, M. *The Sage Handbook of Leadership*. Thousand Oaks, CA: SAGE Publications, Inc., 2011.

Bullis, C. and Reed, G. U.S Army War College. *Assessing Leaders to Establish and Maintain Positive Command Climate.* Department of Command, Leadership and Management Study, i-71. Carlisle, PA: Carlisle Barracks Printing, 2003.

Business Dictionary Online. "Values." *Business Dictionary Online.* 2012. Accessed 10 April 2017 from http://www.businessdictionary.com/definition/values.html.

Cameron, K. "Responsible Leadership as Virtuous Leadership." *Journal of Business Ethics* 98 no. S1 (2011): 25-35.

Ciulla, J. *The Ethics of Leadership.* Belmont, CA: Wadsworth/Thompson Learning, 2003.

Ciulla, J. and Forsyth, D. "Leadership Ethics." In *The Sage Handbook of Leadership,* edited by A. Bryman, D. Collinson, K. Grint, B. Jackson, and M. Uhl-Bien, 229-241. London: SAGE Publications, Ltd., 2011.

Coleman, S. *Military Ethics, an Introduction with Case Studies.* New York, NY: Oxford University Press, 2013.

Conway, J. *Report to Congress on the Posture of the United States Marine Corps.* Headquarters Marine Corps, Washington, D.C., 2010.

Conway, J. "Commanders' Responsibilities in Addressing Sexual Assault." White Letter No. 02-09. Headquarters Marine Corps, Washington, D.C., 2009.

Conway, J. "Contacts with Representatives from Industry." White Letter No. 08-07. Headquarters Marine Corps, Washington, D.C., 2007a.

Conway, J. "34th Commandant of the Marine Corps: Commandant's Planning Guidance." *Leatherneck Magazine,* 90 no. 1 (2007b): 12-25.

Couch, K. (2016). "Officership and Leadership." *Marine Corps Gazette* 100 no. 6 (2016): 22-26.

Cullen, J., Victor, B., and Stephens, C. "An Ethical Weather Report: Assessing the Organization's Ethical Climate." *Organizational Dynamics* 18 no. 2 (1989): 50-62.

Department of the Army, Training, and Doctrine Command. *Field Manual, Army Leadership, no. 6-22* Washington, DC: U.S. Government Printing Office, 2012.

Department of Defense, United States Joint Forces Command. *The Joint Operating Environment (JOE).* Joint Futures Group (J59). Suffolk, VA: Joint Forces Command Printing Office, 2010.

Department of Defense, United States Marine Corps. *Warfighting.* New York, NY: Currency and Doubleday, Dell Publishing Group, Inc., 1989.

Department of the Navy, Headquarters Marine Corps. *Marine Corps Warfighting Publication 6-11, Leading Marines.* Washington, DC: U.S. Government Printing Office, 2014.

Department of the Navy, Headquarters Marine Corps. *Marine Corps Reference Publication 6-11B, Marine Corps Values.* Washington, DC: U.S. Government Printing Office, 2008.

Department of the Navy, Headquarters Marine Corps. *Marine Corps Reference Publication 6-11, Marine Corps Values: A User's Guide for Discussion Leaders.* Washington, DC: U.S. Government Printing Office, 2014.

Department of the Navy, Headquarters Marine Corps. *Marine Corps Reference Publication 6-11D, Sustaining the Transformation.* Washington, DC: U.S. Government Printing Office, 2014.

Department of the Navy, Headquarters Marine Corps, History Division. *Marine Corps Command Chronology.* Quantico, VA: U.S. Government Printing Office, 1987-1991.

Department of the Navy, Headquarters Marine Corps, History Division. *Marine Corps Command Chronology.* Quantico, VA: U.S. Government Printing Office, 1995-1999.

Department of the Navy, Headquarters Marine Corps, History Division. *Marine Corps Command Chronology.* Quantico, VA: U.S. Government Printing Office, 1999-2003.

Department of the Navy, Headquarters Marine Corps, History Division. *Marine Corps Command Chronology.* Quantico, VA: U.S. Government Printing Office, 2003-2007.

Department of the Navy, Headquarters Marine Corps, History Division. *Marine Corps Command Chronology.* Quantico, VA: U.S. Government Printing Office, 2006-2010.

Department of the Navy, Headquarters Marine Corps. *United States Marine Corps Concepts and Programs 2009.* Washington, D.C: U.S. Government Printing Office, 2009.

Department of the Navy, Headquarters Marine Corps. *The Marine Corp Manual.* Washington, D.C: U.S. Government Printing Office, 1982.

Department of the Navy, Headquarters Marine Corps. *Marine Corps Strategy 21.* Washington, D.C: U.S. Government Printing Office, 2000.

Department of the Navy, Headquarters Marine Corps. *Marine Corps Vision and Strategy 2025, Implementation Planning Guidance,* Washington, D.C: U.S. Government Printing Office, 2008.

Department of the Navy, Headquarters Marine Corps. *The 31st Commandant's Planning Guidance.* Quantico, VA: The Marine Corps Association, 1995.

Department of the Navy, Headquarters Marine Corps. *35th Commandant's Planning Guidance.* Quantico, VA: HQMC Printing Office, 2010.

Department of Defense. *Field Manual (FM) 22-100, Army Leadership.* Washington, DC: Government Printing Office, 1999.

Department of the Navy, Headquarters Marine Corps. *Strategic Communication Plan.* Washington, D.C: Government Printing Office, 2007.

Department of the Navy, Headquarters Marine Corps. *Marine Corps Warfighting Publication 6-11, Leading Marines.* Washington, DC: Government Printing Office, 2014.

Department of the Navy, Headquarters Marine Corps. *Marine Corps Order 1500.56: Marine Corps Values Program.* Washington, DC: Government Printing Office, 1996.

Department of the Navy. *Marine Corps Warfighting Publication 5-1, Marine Corps Planning Process.* Washington, DC: Government Printing Office, 2010.

Department of the Navy. *Joint Guided Discussion Playbook.* Core Beliefs Lesson Plan. Marine Corps Recruit Depot, San Diego. Government Printing, 2015.

Department of the Navy, *Joint Guided Discussion Playbook.* Introduction to Ethics Lesson Plan. Marine Corps Recruit Depot, San Diego, CA. Government Printing, 2016.

Department of the Navy. *Ethics I, II and Combat Ethics Discussion.* Student Handout, The Basic School, Marine Corps Training Command, Camp Barrett, Virginia. Government Printing, 2016.

Department of the Navy, Headquarters Marine Corps. *The 36th Commandant's Planning Guidance.* Washington, D.C, 2015b.

Deshpande, S. "Ethical Climate and the Link Between Success and Ethical Behavior: An Empirical Investigation of a Non-Profit Organization.: *Journal of Business Ethics* 15 no. 3 (1996): 315-320.

Deshpande, S., George, E., and Joseph, J. "Ethical Climates and Managerial Success in Russian Organizations." *Journal of Business Ethics* 23 no. 2 (2000): 211-217.

Deshpande, S., Joseph, P., and Shu, X. "Ethical Climate and Managerial Success in China." *Journal of Business Ethics* 99 no. 4 (2011): 527-534.

Donmoyer, R. "Generalizability and the Single-Case Study." In *Qualitative Research in Education: The Debate Continues,* Ed. by E. Eisner and A. Peshkin, 25. New York, NY: Teachers College Press, 1990.

Doty, J., and Doty, C. "Command Responsibility and Accountability." *Military Review* (2012 Jan-Feb): 35-38.

Doty, J., and Gelineau, J. "Command Climate." *Army Magazine,* July 2008, 22-24.

Dunford, J. *Brief to the House Appropriations Committee Subcommittee on Defense.* Headquarters Marine Corps, Washington, D.C., 2015a.

Erhart, M., Schneider, B., and Macey, W. *Organizational Climate and Culture: An Introduction to Theory, Research, and Practice.* New York, NY: Routledge, 2014.

Ficarrotta, J. *Kantian Thinking about Military Ethics.* Burlington, VT: Ashgate Publishing, 2010.

Flynn, G. "Keepers of the Flame." *Joint Forces Quarterly, 3d Quarter* 54 (2009): 16-19.

Forester, C.S. *Rifleman Dodd.* Baltimore, MD: The Nautical and Aviation Publishing Company of America, 1989.

Fowler, F.J. *Survey Research Methods.* Thousand Oaks, CA: Sage Publications, 2014.

Freedman, D. *Corps Business: The 30 Management Principles of the U.S. Marines.* New York, NY: Harper Collins Publishers, 2000.

Gray, A. and Otte, P. *The Conflicted Leader and Vantage Leadership.* Columbus, OH: Franklin University Press, 2006.

Grojean, M., Resick, C., Dickson, M., and Smith, D. "Leaders, Values, and Organizational Climate: Examining Leadership Strategies for Establishing an Organizational Climate Regarding Ethics." *Journal of Business Ethics* 55 no. 3 (2004): 223-241.

Glesne, C. *Becoming Qualitative Researchers, An Introduction*, New York, NY: Pearson, 2016.

Harkins, G., and Schogol, J. "Marines find culture of hazing, abuse at boot camp after recruit's death at Parris Island." *Marine Corps Times*, September 8, 2016. Retrieved from marines-find-culture-of-hazing-abuse-at-boot-camp-after-recruit-s-death-at-parris-island/.

Haslam, S., Reicher, S., and Platow, M. *The New Psychology of Leadership: Identity, Influence, and Power.* New York, NY: Psychology Press, 2011.

Heifetz, R. *Leadership Without Easy Answers.* Cambridge, Massachusetts: Harvard University Press, 1994.

Heifetz, R., and Linsky, M. *Leadership on the Line: Staying Alive through the Dangers of Leading.* Boston, Massachusetts: *Harvard Business Review* Press, 2002.

Hinman, L. *Ethics, A Pluralistic Approach to Moral Theory.* Boston, MA: Wadsworth, 2013.

Hipple, J., and Olson, S. "Values-Based Leadership: Lessons from the United States Marine Corps." *Harvard Business Review* (2010): 1-24.

Horn, W. "A Leader's Tool." *Marine Corps Gazette* 100 no. 9 (2016): 37-39.

Immel, A. "The Need for an Ethical Fitness Assessment in the US Armed Forces." *Journal of Military Ethics* 15 no. 1 (2016): 3-17.

Johnson, C. *Meeting the Ethical Challenges of Leadership, Casting Light or Shadow.* Thousand Oaks, CA: SAGE Publications, Inc, 2015.

Johnson, R. "How Can Leaders Maintain Ethical Command Climates?" *LTX Magazine* 3 no. 2 (2013): 1-3.

Kahneman, D. *Thinking Fast and Slow.* New York, NY: Farrar, Straus, and Giroux, 2011.

Katolin, D. "Ethical Development: How We Shape Our People's Behavior." *Marine Corps Gazette* 100 no. 9 (2016a): 53-57.

Katolin, D. "Military Ethics: How Maneuver Warfare Demands Adherence to Our Values." *Marine Corps Gazette* 100 no. 8 (2016b): 44-46.

Keenan, J. "Protect What You've Earned: Engaging in the Conversation." *Marine Corps Gazette* 101 no. 3 (2017): 60-62.

Kerr, J. *Legacy: What the All Blacks Can Teach Us about Life.* London, UK: Constable and Robinson Ltd, 2013.

Kirkpatrick, S. A., and Locke. E.A. "Leadership: Do traits matter?" *Academy of Management Executive* 5 no. 2 (1991): 48–60.

Kuenzi, M. and Schminke, M. "Assembling Fragments into a Lens: A Review, Critique, and Proposed Research Agenda for the Organizational Work Climate Literature." *Journal of Management* 35 no. 3 (2009): 634-717.

Krulak, C. *Professional Reading Program; Commandant's Choice for 1995*. ALMAR 157/95, 1995.

Krulak, C. "Character." *Leatherneck Magazine,* 79 no. 12 (1996a): 14-15.

Krulak, C. "Marine Corps Values Program." Marine Corps Order 1500.56. Headquarters Marine Corps. Washington: D.C. Government Printing, 1996b.

Krulak, C. "Marine Corps Values Program." White Letter No. 16-96 (1996c). Retrieved from http://www.usmc.mil/cmcalmars.

Krulak, C. *Implementing Instructions for the Marine Corps Values Program.* ALMAR 439/96, 1996d.

Krulak, C. *Report on the Progress of the Marine Corps Values Program.* ALMAR 238/97, 1997.

Krulak, C. "Sustaining the Transformation." White Letter No. 3-98. (1998).

Krulak, V. *First to Fight: An Inside View of the U.S. Marine Corps.* Annapolis, MD: Naval Institute Press, 1984.

Lenhardt, T. "Leading Marines: A Compassion Strengthened by Emotional Stability." *Marine Corps Gazette* 100 no. 6 (2016): 42-45.

Lieutenants, 6th Platoon, Delta Company. "The Warrior's Leader: Promoting Discipline Through Culture." *Marine Corps Gazette* 100 no. 7 (2016): 29-30.

Lejeune, John A. *Reminiscences of a Marine.* University of Wisconsin-Madison, WI: Arno Press, 1930.

Lucas, G. *Routledge Handbook of Military Ethics.* New York, NY: Routledge, 2015.

Maak, T., and Pless, N. "Understanding Responsible Leadership: Role Identity and Motivational Drivers." *Journal of Business Ethics,* 74 no. 4 (2007): 437-456.

Major, A. E. "Ethics Education of Military Leaders." *Military Review* (2014 Mar-Apr): 55-60.

Major, E., DeRemer, L., and Bolgiano, D. "Ethics Can Be Taught." *Proceedings* 138 no. 12 (2012): 1-5. Retrieved from http://www.usni.org.

Marine Corps University. *Proceedings from the Russell Leadership Conference 2008.* After Action Report, Quantico, VA, 2008.

Marino, G. *Ethics, The Essential Writings.* New York, NY: Random House Inc., 2010.

Martin, K., and Cullen, J. "Continuities and Extensions of Ethical Climate Theory: A Meta-Analytic Review." *Journal of Business Ethics* 69 no. 2 (2006): 175-194.

Mayer, D., Kuenzi, M., and Greenbuam, R. "Examining the Link between Ethical Leadership and Employee Misconduct: The Mediating Role of Ethical Climate." *Journal of Business Ethics* 95 no. S1 (2010): 7-16.

Mayer, D. "A Review of the Literature on Organizational Climate and Culture." *The Oxford Handbook of Organizational Climate and Culture* (2014): 1-28. Retrieved from http://www.oxfordhandbooks.com.

McAleer, P. "Our Corps' Soul at Stake." *Marine Corps Gazette,* 101 no. 6 (2017): 47-48.

Merriam, S. and Tisdell, E. *Qualitative Research, A Guide to Design and Implementation.* San Francisco, CA: Jossey-Bass, 2016.

Miller, C., and Poole, E. *Ethical Leadership, Global Challenges and Perspectives.* New York, NY: Palgrave MacMillan, 2011.

Montor, K. *Ethics for the Junior Officer.* Annapolis, MD: Naval Institute Press, 2001.

Myer, A., Thoroughgood, C., and Mohammed, S. "Complementary or Competing Climates? Examining the Interactive Effect of Service and Ethical Climates on Company-Level Financial Performance." *Journal of Applied Psychology* 15 no. 3 (2016): 1178-90.

Neller, R. "A Message from the Commandant of the Marine Corps." 241st Birthday of the Marine Corps. Headquarters Marine Corps, Washington, D.C, 2016.

Neller, R. "Learning Without Losing." *The Marine Corps Gazette* 101 no. 11 (2017a): 6-8.

Neller, R. "Recent Conduct on Social Media." White Letter No. 1-17. Headquarters Marine Corps, Washington, D.C, 2017b.

Neller, R. "Social Media Guidance—Mandatory Counseling Requirement." White Letter No. 2-17. Headquarters Marine Corps, Washington, D.C, 2017c.

Neller, R. *The Commandant's Posture of the United States Marine Corps, President's Budget 2017*. (2017d.) Retrieved from www.hqmc.marines.mil/.../cmc%20PB17%20Posture%20written%20Testimony%20Final.pdf on 17 May 2018.

Olsthoorn, P. *Military Ethics and Virtues*. New York, NY: Routledge, 2011.

Otte, P. *Grayisms*. Arlington, VA: Potomac Institute Press, 2015.

Patton, M. *Qualitative Research, and Evaluation Methods*. Thousand Oaks, CA: SAGE Publications Ltd., 2015.

Peshkin, A. "In Search of Subjectivity-One's Own." *Educational Researcher* (October 1988): 17-21.

Pfeffer, J., Salancik, G., and Leblebici, H. *The Effect of Uncertainty on the Use of Social Influence in Organizational Decision Making*. Lexington, KY: The University Press of Kentucky, 1976.

Polkinghorne, D. "Narrative Configuration in Qualitative Analysis." *Qualitative Studies in Education* 8 no. 1 (1995): 5-23.

Potter, E.B. *Nimitz*. Annapolis, MD: Naval Institute Press, 1976.

Robinson, P. "Ethics Training and Development in the Military." *Parameters* 37 no. 1 (2007): 23-36.

Robinson, P., De Lee, N., and Carrick, D. *Ethics Education in the Military*. Burlington, VT: Ashgate Publishing, 2008.

Rowell IV, G. *Marine Corps Values-Based Ethics Training: A Recipe to Reduce Misconduct, Strategy Research Project, 1-29)*. United States Army War College. Carlisle, PA: Carlisle Barracks Printing, 2013.

Saldana, J. *The Coding Manual for Qualitative Researchers.* Thousand Oaks, CA: Sage Publications, Inc., 2016.

San Diego Union Tribune. "Marine Corps Investigating Nude Photo Scandal." *San Diego Union Tribune,* March 7 2017a. Retrieved from http://www.sandiegouniontribune.com.

San Diego Union Tribune. "9 More Navy Officers Indicted." *San Diego Union Tribune,* March 15 2017b. Retrieved from http://www.sandiegouniontribune.com.

Schaubroeck, J., Hannah, S., Avolio, B., Kozlowski, S., Lord, R., Trevino, L., Dimotakis, N., and Peng, A. "Embedding Ethical Leadership Within and Across Organization Levels." *Academy of Management Journal* 55 no. 5 (2012): 1053-1078.

Schein, E. *Organizational Culture and Leadership.* San Francisco, CA: Jossey-Bass, 2010.

Schminke, M., Arnaud, A., and Kuenzi, M. "The Power of Ethical Work Climates." *Organizational Dynamics* 36 no. 2 (2007): 171-186.

Schneider, B., Ehrhart, M., and Macey, W. "Organizational Climate Research: Achievements and the Road Ahead." In *Handbook of Organizational Culture and Climate,* 2nd *edition*, edited by N.M. Ashkanasay, C.P.M. Wilderom, and M.F. Peterson (29-49). Thousand Oaks, CA: Sage, 2011a.

Shisler, G. *For Country and Corps, The Life of General Oliver P. Smith.* Annapolis, MD: Naval Institute Press, 2009.

Simha, A., and Cullen, J. "Ethical Climates and Organizational Outcomes: Implications from the Past and Prophecies for the Future." *Academy of Management Perspectives (Nov 2012):* 20-34.

Sledge, E.B. *With the Old Breed.* New York, NY: Presidio Press, 2007.

Socratic Method. Retrieved from http://www.yourdictionary.com/socratic.method 15 May 2018.

United States Code. *Title 10, National Security Act.* (1947) Retrieved from https://en.wikipedia.org/wiki/National_Security_Act_of_1947.

Vaughn, L. *Doing Ethics, Moral Reasoning and Contemporary Issues.* New York, NY: W.W. Norton and Company, Inc., 2013.

Victor, B., & Cullen, J. B. (1987). A theory and measure of ethical climate in organizations. Research in Corporate Social Performance and Policy, 9, 51-71.

Victor, B., & Cullen, J. B. (1988). The organizational bases of ethical work climates. Administrative Science Quarterly, 33, 101-125.

Weber, J., and Gerde, V. "Organizational Role and Environmental Uncertainty as Influences on Ethical Climate in Military Units." *Journal of Business Ethics* 100 (2010): 595-612.

Whiffen, H. "Becoming an Adaptive Leader." *Military Review, Special Edition: The Profession of Arms* (Sept 2007): 92-98.

Wong, L.. (Director, Strategic Studies Institute, U.S. Army War College.) *Developing Adaptive Leaders: The Crucible Experience of Operation Iraqi Freedom.* Carlisle, PA, 2004.

Wortel, E. and Bosch, J. "Strengthening Moral Competency: A Train the Trainer Course on Military Ethics." *Journal of Military Ethics* 10 no. 1 (2011): 17-35.

Yin, R. *Case Study Research: Design and Methods.* Thousand Oaks, CA: SAGE, 2003.

Index

U

V

W

Z

Made in the USA
Monee, IL
05 November 2026

67811459R00139